1984

ALLEN H. MACKENZIE, M. D.

Low-Dose Methotrexate Therapy in Rheumatic Diseases

Rheumatology

An Annual Review

Vol. 9

Series Editor
M. Schattenkirchner, München

KARGER

Basel · München · Paris · London · New York · New Delhi · Singapore · Tokyo · Sydney

Symposium on Low-Dose Methotrexate Therapy in Rheumatic Diseases,
Schloss Hugenpoet/Düsseldorf, September 28–30, 1984

Low-Dose Methotrexate Therapy in Rheumatic Diseases

Volume Editor
R. Rau, Ratingen

53 figures, 1 color plate, 59 tables, 1986

KARGER

Basel · München · Paris · London · New York · New Delhi · Singapore · Tokyo · Sydney

National Library of Medicine, Cataloging in Publication
Symposium on Low-Dose Methotrexate Therapy in Rheumatic Diseases
(1984: Düsseldorf, Germany)
Low-Dose methotrexate therapy in rheumatic diseases/
Symposium on Low-Dose Methotrexate Therapy in Rheumatic Diseases,
Schloss Hugenpoet/Düsseldorf, September 28–30, 1984;
Volume editor, R. Rau. – Basel; New York: Karger, 1986.
(Rheumatology, an annual review; vol. 9)
Dedicated to H.G. Lasch on the occasion of his 60th birthday.
Includes index.
1. Methotrexate – administration & dosage – congresses 2. Methotrexate – therapeutic use – congresses 3. Rheumatism – drug therapie – congresses
I. Rau, Rolf, 1933 – II. Lasch, Hanns-Gotthard III. Title IV. Series
W1 RH42G v.9
WE 544 S9888L 1984

Drug Dosage

The authors and the publisher have exerted every effort to ensure that drug selection and dosage set forth in this text are in accord with current recommendations and practice at the time of publication. However, in view of ongoing research, changes in government regulations, and the constant flow of information relating to drug therapy and drug reactions, the reader is urged to check the package insert for each drug for any change in indications and dosage and for added warnings and precautions. This is particularly important when the recommended agent is a new and/or infrequently employed drug.

Printed in Switzerland by gdz (Genossenschaftsdruckerei Zürich)
ISBN 3–8055–4236–4

Contents

List of Participants

Dunky, A., Dr. med., 2. Medizinische Abteilung, Krankenhaus der Stadt Wien-Lainz, Wolkersbergenstrasse 1, A–1130 Wien (Österreich)

Fenner, H., Prof., Institut für Pharmazie der Freien Universität Berlin, Königin-Luise-Strasse 2–4, D–1000 Berlin 33

Frenzel, H., Prof. Dr. med., Pathologisches Institut der Universität Düsseldorf, Moorenstrasse 5, D–4000 Düsseldorf (BRD)

Furst, D.E., MD, Associate Professor, Division of Rheumatology, Department of Internal Medicine, University of Iowa, East 400 th, Iowa City, IA 52242 (USA)

Gross, D., Prof. Dr. med., Klinik für Rheumatologie und Rehabilitation, Stadtspital Triemli, Birmensdorferstrasse, CH–8063 Zürich (Schweiz)

Herborn, G., Dr. med., Rheumaklinik, Evangelisches Fachkrankenhaus, Rosenstrasse 2, D–4030 Ratingen (BRD)

Jüptner, J., Dr. med., Cyanamid GmbH, Lederle Arzneimittel, Pfaffenrieder Strasse 7, D–8190 Wolfratshausen (BRD)

Jürgens, H., PD Dr. med., Kinderklinik der Universität Düsseldorf, Moorenstrasse 5, D–4000 Düsseldorf (BRD)

Karger, T., Dr. med., Rheumaklinik, Evangelisches Fachkrankenhaus, Rosenstrasse 2, D–4030 Ratingen (BRD)

Katzenberger, K., Dr. med., Cyanamid GmbH, Lederle Arzneimittel, Pfaffenrieder Strasse 7, D–8190 Wolfratshausen (BRD)

Kiltie, H., MD, Medical Research, International Lederle Laboratories, Pearl River, NJ 10965 (USA)

Kriegel, W., PD Dr., St. Willibrordus-Spital, Willibrordusstrasse 9, D–4240 Emmerich (BRD)

Mackenzie, A.H., MD, Head, Section on Clinical Rheumatology, Department of Rheumatic and Immunologic Diseases, The Cleveland Clinic Foundation, 9500 Euclid Avenue, Cleveland, OH 44106 (USA)

Michaels, R.M., MD, FACP, Division of Rheumatology, Guthrie Clinic Ltd., Guthrie Square, Sayre PA 18840 (USA)

Miescher, P.A., Prof. Dr. med., Division d'Hématologie et Centre de Transfusion Sanguine, 25, rue Micheli-du-Crest, CH–1211 Genève (Suisse)

Mitrenga, D., Dr., Krankenhaus der Augustinerinnen, Jacobstrasse 27–31, D–5000 Köln 1 (BRD)

Nyfors, A., Prof. Dr. med., Department of Dermatology, University of Bergen, Haukeland Sykehus, N–5016 Bergen (Norway)

Pöhler, E., Dr. med., Medizinische Klinik der Universität, Joseph-Stelzmann-Strasse 9, D–5000 Köln 41 (BRD)

Rau, R., PD Dr. med., Rheumaklinik, Evangelisches Fachkrankenhaus, Rosenstrasse 2, D–4030 Ratingen (BRD)

Sauer, H., Prof. Dr. med., Medizinische Klinik III, Klinikum Grosshadern, Ludwig-Maximilians-Universität, Marchioninistrasse 15, D–8000 München 70 (BRD)

Schilling, F., Prof. Dr., Hebbelstrasse 20, D–6500 Mainz 31 (BRD)

Schmidt, K.L., Prof. Dr., Klinik für Physikalische Medizin, Balneologie und Rheumatologie der Universität Giessen, Ludwigstrasse 37–39, D–6350 Bad Nauheim (BRD)

Stroehmann, I., Prof. Dr., Röntgenstrasse 6, D–5300 Bonn 2 (BRD)

Warnatz, H., Prof. Dr., Katholisches Krankenhaus St. Josef, Brückstrasse 91–93, D–4300 Essen 16 (BRD)

Weinstein, A., MD, Asst Prof., Division of Rheumatic Diseases and Immunology, New York Medical College, Valhalla, NY 10595 (USA)

Wilke, W.S., MD, Department of Rheumatic and Immunologic Diseases, The Cleveland Clinic Foundation, 9500 Euclid Avenue, Cleveland, OH 44106 (USA)

Williams, H.J., MD, University of Utah Medical Center, 50 North Medical Drive, Salt Lake City, UT 84132 (USA)

Willkens, R.F., MD, Clinical Professor of Medicine, Harborview Medical Center, University of Washington, 325 Ninth Avenue, Seattle, WA 98104 (USA)

Zeidler, H., Prof. Dr., Medizinische Klinik und Poliklinik der Universität Düsseldorf, Moorenstrasse 5, D–4000 Düsseldorf (BRD)

Foreword

The valuation of drugs used in the therapy of rheumatic disorders fluctuates considerably from the time of introduction until a final decision is made. This, especially, has occurred with the so-called remission-inducing agents in rheumatoid arthritis, such as gold compounds, chloroquine, and penicillamine, as well as with some of the immunosuppressive or cytotoxic substances.

The selection of those possibly disease-modifying antirheumatic drugs is small compared with the large number of analgesic and anti-inflammatory substances. A breakthrough in the sense of a curative treatment of rheumatic diseases is not to be expected in the near future.

Therefore, a thorough evaluation of the most interesting pharmacological and clinical aspects of methotrexate is very useful. The antiarthritic potency of methotrexate was detected 35 years ago. During the last years, low-dose oral therapy of methotrexate turned out to be an enrichment to the therapy of rheumatoid arthritis. In a low-dose regimen, the hepatotoxicity of methotrexate seems to be a minor problem. The risk of causing a malignancy also seems to be neglectible.

R. Rau, known as an expert in the field of rheumatic diseases and liver problems undertook the task of organizing a convention of the most famous experts in methotrexate therapy. The proceedings of this international workshop submitted in this volume will certainly be well accepted both by clinicians and basic research workers in rheumatology.

M. Schattenkirchner

Preface

Some 35 years ago, shortly after its introduction as an antineoplastic agent, the efficacy of methotrexate in the treatment of rheumatoid arthritis and psoriasis was demonstrated.

While the treatment of otherwise refractory psoriasis with low doses of methotrexate was developed during the following years and decades, low-dose methotrexate therapy of resistant forms of rheumatoid arthritis was increasingly employed only during the last four years, based on publications which appeared since 1980.

This volume comprises the papers and discussions from a symposium which took place in September 1984. Rheumatologists having the most extensive experience with this therapy were invited together with experts from the fields of pharmacology, oncology, immunology, dermatology, pathology and pulmonology, in order to give a review of our current knowledge, and also to discuss considerations on indications, efficacy, and possible side effects of this form of therapy. I would like to thank my colleagues, Dr. *T. Karger* and Dr. *G. Herborn,* for their cooperation in preparing the congress.

My special thanks go to the German and the American Cyanimid Company, and in particular to Dr. *K. Katzenberger,* for their optimal support, both organizational and financial, in the preparation and realization of the congress and in the publication of this volume.

R. Rau

This volume is dedicated to
Prof. Dr. Dr. h.c. *H.G. Lasch*
on the occasion of his 60th birthday

Rheumatology, vol. 9, pp. 1–5 (Karger, Basel 1986)

Introduction: Short History of Methotrexate Therapy in Psoriasis and Rheumatic Diseases

R. Rau

Department of Rheumatology, Evangelisches Fachkrankenhaus, Ratingen, FRG

In the words of *Rees* [1], the history of methotrexate (MTX) treatment is as fascinating as the discovery of cortisone. In 1946, folic acid was identified and synthesized by a research team of the American Cyanamid Company [2]. Two years earlier an antiproliferative effect of splenic extracts containing folic acid analogues in animal experiments had been described by *Leuchtenberger* et al. [3]. *Farber* et al. [4] tried to treat 11 patients with acute leukemia by the means of such extracts but these folic acid analogues aggravated the course of the leukemia. This experience was the reason to search for compounds possessing antifolic acid properties.

In 1947, *Franklin* et al. [5] demonstrated that x-methylfolic acid inhibited the growth of rats and depressed their blood counts and bone marrow. In the same year, *Seeger* et al. [6] synthesized the folic acid antagonist aminopterine. Already in 1948, *Farber* et al. [7] succeeded in obtaining temporary remission of leukemia in 5 children by daily injections of 0.5–1.0 mg aminopterine. In search for a less toxic agent than aminopterine, *Smith* et al. [8] in 1948 synthesized MTX (4-amino-10-methyl folic acid).

Although aminopterine – and later MTX – was widely used only as an oncologic agent, already in 1951 *Gubner* et al. [9] published an article on the treatment of rheumatoid arthritis, a nonmalignant disease, with aminopterine. This was one year after the report of *Jimenez-Diaz* [9a] on the effect of nitrogen mustard on rheumatoid arthritis. They studied 6 patients with rheumatoid arthritis, 1 with psoriatic arthritis and 1 with rheumatic fever, who were treated orally with a daily dose of 1–2 mg aminopterine. While 5 out of their 6 rheumatoid patients improved rather quickly, in the patient with psoriasis arthritis, psoriasis cleared rapidly. Although the report concerning efficacy in rheumatoid arthritis was more or less forgotten, *Rees* and co-workers [10, 11] adopted the treatment for psoriasis in daily doses of 0.5–1.0 mg aminopterine and in publications from 1954 and 1959 reported on clearing or great improvement of psoriasis in 80% of 171 and 229 patients, respectively [10, 11]. In the following year, MTX in daily doses of 2.5 mg was found to be as effective but less toxic than aminopterine [12, 13].

After animal experiments had shown that single high doses of MTX were less toxic than several low doses of the same total quantity [14, 15], and that the toxicity was directly proportional to the duration of exposure [16], *Roenigk* et al. [17] in 1965 recommended one single weekly oral dose of 25 mg MTX. In accord with the 37-hour cell cycle of the basal cell layer in psoriasis, *Weinstein und Frost* [18] in 1971 recommended administering the weekly oral dose of MTX in three portions at 12-hour intervals.

Hepatic fibrosis and cirrhosis proved more common in psoriatics treated with daily oral doses than in patients treated with weekly oral doses [19], and more common in patients who had received MTX more than 12 times a month than in patients with fewer monthly doses [20]. The frequency of liver fibrosis was higher in patients who were treated with three divided doses of MTX (33 patients) than in patients with one weekly dose (43 patients) [21]. In the opinion of *Nyfors* [22], who has a vast experience in the treatment of psoriasis with MTX and who made a large number of liver biopsies, the three-divided dose was a good idea in respect of cell kinetics in the epidermis, but appears unfortunate from the pharmacokinetic and hepatotoxic point of view. Nevertheless, the regimen of *Weinstein and Frost* [18] is still followed by most of the dermatologists and rheumatologists.

During the 1960s MTX was extensively used in the treatment of psoriasis, resistant to other forms of therapy. Although the drug was not approved by the FDA for this indication until 1972, *Weinstein* estimated that as many as 15,000 patients with refractory psoriasis were being clinically treated with MTX in the year 1971. The efficacy of MTX in psoriasis arthritis was demonstrated by *Black* et al. [23] in 1964 and since then seems to be widely used for this indication.

According to *Willkens* et al. [29], *Hoffmeister's* [24] publication in 1972 was one of the first regarding the treatment of rheumatoid arthritis, but already 5 years earlier, in 1967, *Gross* and co-workers [25, 26] reported on 31 patients with rheumatoid arthritis treated with one weekly intravenous injection of 50 mg MTX with improvement of the clinical condition and the sedimentation rate. They followed the recommendations of *Miescher and Riethmüller* [27] who in 1964 in Geneva and New York, respectively, began to treat collagen vascular diseases, especially systemic lupus erythematosus, with MTX after having obtained good results in the treatment of experimental thyroiditis in guinea pigs [28]. In the maintenance therapy they administered MTX and azathioprine (or 6-mercaptopurine) on an alternate monthly schedule.

Although some American rheumatologists – for instance the Cleveland Clinic – seem to have continued treating their patients with MTX, until 1980 there were almost no reports on MTX for rheumatic diseases. Since 1980 a new and greater interest in MTX treatment of rheumatic diseases can be observed and is documented in reports on the treatment of rheumatoid arthritis by *Willkens* et al. [29, 31], *Wilke* et al. [30, 33], *Michaels* et al. [32] and *Ward* [34], of Reiter's syndrome [35, 36] and of psoriatic arthritis [37]. Our own experience dates back to 1979, when we also started to treat patients with rheumatoid arthritis [38] after having obtained good results in the treatment of psoriatic arthritis and Reiter's syndrome. As most of the above mentioned investigators could be invited to contribute articles and discussions to this volume, the reader will get comprehensive and complete information about the current knowledge and the current problems of MTX treatment in rheumatic diseases. The major problem of this therapy – if long-term treatment with MTX can induce chronic liver disease – is discussed extensively in this book.

References

1 Rees, R.B.: Methotrexate: 20 years later; in Farber, Cox, Proc. 2nd Int. Symp. on Psoriasis, pp. 231–234 (Yorke Medical Books, New York 1977).

2 Angier, R.B.; Boothe, J.H.; Hutchings, B.L.; Mowat, J.H.; Semb, J.; Stokstad, E.L.R.; Subbarow, Y.; Waller, C.W.: The structure and synthesis of the liver *L. casei* factor. Science *103:* 667–669 (1946).

3 Leuchtenberger, R.; Leuchtenberger, C.; Laszlo, D.; Lewisohn, R.: The influence of 'folic acid' on spontaneous breast cancers in mice. Science *101:* 46 (1945).

4 Farber, S.; Cutler, E.C.; Hawkins, J.W.; Harrison, J.H.; Peirce, E.C.; Lenz, G.G.: The action of pteroylglutamic conjugates on man. Science *106:* 619–621 (1947).

5 Franklin, A.L.; Stokstad, E.L.R.; Belt, M.; Jukes, T.H.: Biochemical experiments with a synthetic preparation having an action antagonistic to that of pteroylglutamic acid. J. biol. Chem. *169:* 427–436 (1947).
6 Seeger, D.R.; Smith, J.M., Jr.; Hultquist, M.E.: Antagonist for pteroylglutamic acid. J. Am. chem. Soc. *69:* 2567 (1947).
7 Farber, S.; Diamond, L.K.; Mercer, R.D.; Sylvester, R.F.; Wolff, J.A.: Temporary remissions in acute leucemia in children produced by folic acid antagonist, 4-amino-pteroyl-glutamic acid (aminopterin). New. Engl. J. Med. *238:* 787–793 (1948).
8 Smith, J.M., Jr.; Cosulich, D.B.; Hultquist, M.E.; Seeger, D.R.: The chemistry of certain pteroylglutamic acid antagonists. Trans. N.Y. Acad. Sci. *10:* 82–83 (1948).
9 Gubner. R; August, S.; Ginsberg, V.: Therapeutic suppression of tissue reactivity. II. Effect of aminopterin in rheumatoid arthritis and psoriasis. Am. J. med. Sci. *221:* 176–182 (1951).
9a Jimenez-Diaz, C.; García, E.; Merchante, A.; Derianes, J.: Treatment of rheumatoid arthritis with nitrogen mustard. J. Am. med. Ass. *147:* 1418 (1951).
10 Rees, R.B.; Bennett, J.H.; Bostick, W.L.: Aminopterin for psoriasis. Archs. Derm. *72:* 133–143 (1955).
11 Rees, R.B.; Bennett, J.H.: Further observations on aminopterin for psoriasis. J. invest. Derm. *32:* 61–66 (1959).
12 Edmundson, W.F.; Guy, W.B.: Treatment of psoriasis with folic acid antagonists. Archs. Derm. *78:* 200–203 (1958).
13 Rees, R.B.; Bennett, J.H.: Methotrexate versus aminopterin for psoriasis. Archs. Derm. *83:* 970–972 (1961).
14 Ferguson, F.C., Jr.; Thiersch, J.B.; Philips, F.S.: The action of 4-amino-N-10-methyl-pteroylglutamic acid in mice, rats and dogs. J. Pharmac. exp. Ther. *98:* 293–299 (1950).
15 Rall, D.; Mead, J.A.R.; Freireich, E.J.; Van Scott, E.J.; Hertz, R.; Lipsett, M.B.: Folic acid antagonists: effects on the cell and the patient. Ann. intern. Med. *59:* 931–956 (1963).
16 Bruce, W.R.; Meeker, B.E.; Valeriote, F.A.: Comparison of the sensitivity of normal hematopoietic and transplanted lymphoma colony-forming cells to chemotherapeutic agents administered in vivo. J. natn. Cancer Inst. *37:* 233–245 (1966).
17 Roenigk, H.H., Jr.; Haserick, J.R.; Curtis, G.H.: Methotrexate for psoriasis. A preliminary report. Cleveland Clin. Q. *32:* 211–215 (1965).
18 Weinstein, G.D.; Frost, P.: Methotrexate for psoriasis. Archs. Derm. *103:* 33–38 (1971).
19 Dahl, M.G.C.; Gregory, M.M.; Scheuer, P.J.: Methotrexate hepatotoxicity in psoriasis – comparison of different dose regimes. Br. med. J. *i:* 654–656 (1972).
20 Podurgiel, B.J.; McGill, D.B.; Ludwig, J.; Taylor, W.F.; Muller, S.A.: Liver injury associated with methotrexate therapy for psoriasis. Mayo Clin. Proc. *48:* 787–792 (1973).
21 Roenigk, H.H., Jr.; Auerbach, R.; Bergfeld, W.F.; Frost, P.; Maibach, H.I.; Tindall, J.; Weinstein, G.D.: A cooperative, prospective study of the effects of chemotherapy of psoriasis on liver biopsies; in Farber, Cox, Proc. 2nd Int. Symp. on Psoriasis, pp. 243–248 (Yorke Medical Books, New York 1977).
22 Nyfors, A.: Methotrexate therapy of psoriasis. Effects and side effects with particular reference to hepatic changes. A survey. Dan. med. Bull *27:* 74–96 (1980).

23 Black, R.L.; O'Brien, W.M.; Van Scott, E.J.; Auerbach, R.; Eisen, A.Z.; Bunim, J.J.: Methotrexate therapy in psoriatic arthritis. Double-blind study on 21 patients. J. Am. med. Ass. *180:* 141–145 (1964).
24 Hoffmeister, R.T.: Methotrexate in rheumatoid arthritis (Abstract). Arthritis Theum. *15:* 114 (1972).
25 Gross, D.; Enderlin, M.; Fehr, K.: Die immunsuppressive Behandlung der progredient chronischen Polyarthritis mit Antimetabolica und Cytostatica. Schweiz. med. Wschr. *97:* 1301 (1967).
26 Gross, D.; Enderlin, M.: Beitrag zur Behandlung der progredient chronischen Polyarthritis mit Antimetaboliten und Cytostatica. Z. Rheumaforsch. *26* (1967).
27 Miescher, P.A.; Riethmüller, G.: Diagnosis and treatment of systemic lupus erythematosus. Semin. Hematol. *2:* 1 (1965).
28 Spiegelberg, H.L.; Miescher, P.A.: The effect of 6-mercaptopurine and methotrexate on experimental immune thyroiditis in guinea pigs. J. exp. Med. *118:* 869 (1963).
29 Willkens, R.F., et al.: Low dose pulse methotrexate in rheumatoid arthritis. J. Rheumatol. *7:* 501–505 (1980).
30 Wilke, W.S.; Calabrese, L.H.; Scherbel, A.L.: Methotrexate in the treatment of rheumatoid arthritis. Pilot study. Cleveland Clin. Q. *47:* 305–309 (1980).
31 Willkens, R.F.; Watson, M.A.: Methotrexate: a perspective of its use in the treatment of rheumatic disease. J. Lab. clin. Med. *100:* 314–321 (1982).
32 Michaels, R.M.; Nashel, D.J.; Leonard, A.; Sliwinski, J.; Derbes, S.J.: Weekly intravenous methotrexate in the treatment of rheumatoid arthritis. Arthritis Rheum. *25:* 339–341 (1982).
33 Groff, G.D.; Shenberger, K.N.; Wilke, W.S.; Taylor, T.H.: Low-dose oral methotrexate in rheumatoid arthritis. An uncontrolled trial and review of the literature. Semin. Arthrit. Rheumatism *12:* 333–347 (1983).
34 Ward, J.R.: For the cooperative systematic studies of rheumatic disease: a prospective randomized controlled trial of low-dose pulse methotrexate in rheumatoid arthritis. Arthritis Rheum. *27:* suppl., p. 40 (1984).
35 Owen, E.; Cohen, M.: Methotrexate in Reiter's disease. Ann. rheum. Dis. *38:* 48 (1979).
36 Luderschmidt, C.; Balda, B.-R.: Morbus Reiter. Erfolgreiche Kombinationschemotherapie mit Methotrexat, Etretinat und Prednisolon. Münch. med. Wschr. *125:* 936 (1983).
37 Willkens, R.F.; Williams, H.J.; Ward, J.R.; Egger, M.J.; Reading, J.C.; Clements, P.J.; Cathcart, E.S.; Samuelson, C.O.; Solsky, M.A.; Kaplan, S.B.; Guttadauria, M.; Halla, J.T.; Weinstein, A.: Randomized, double-blind, placebo controlled trial of low-dose pulse methotrexate in psoriatic arthritis. Arthritis Rheum. *27:* 376 (1984).
38 Karger, T.; Rau, R.: Methotrexat-Behandlung hochaktiver chronischer Polyarthritiden. Münch. med. Wschr. *126:* 560–564 (1984).

PD Dr. med. R. Rau, Department of Rheumatology, Evangelisches Fachkrankenhaus, Rosenstrasse 2, D-4030 Ratingen (FRG)

Rheumatology, vol. 9, pp. 6–19 (Karger, Basel 1986)

Pharmacology, Toxicology and Drug Interactions of Methotrexate[1]

Herbert Jürgens, Ulrich Göbel

Zentrum für Kinderheilkunde der Universität Düsseldorf, Abteilung für Hämatologie und Onkologie (Leiter: Prof. Dr. *U. Göbel*), Düsseldorf, BRD

For cancer chemotherapy various agents are used with different mechanisms of action, different efficacy and different patterns of toxicity. Amongst the first chemotherapeutic agents studied were folic acid antagonists. The antiproliferative activity in animal tumors was first described by *Leuchtenberger* et al. [1] in 1944. Aminopterin, the classic folic acid antagonist and precursor of methotrexate (amethopterin) (MTX) was first synthesized by *Seeger* et al. [2] in 1947 and clinically used by *Farber* et al. [3] in 1948 when he obtained remissions in ALL with a daily dose of about 2 mg orally. One year later MTX was synthesized by *Smith,* Calco Chemical Division, American Cynamid Company, and rapidly replaced aminopterin in clinical trials on account of its milder pattern of toxicity [4].

A major breakthrough in cancer treatment was the successful use of MTX for the curative treatment of chorionepithelioma as first described by *Li* et al. [5] in 1956. *Murphy* [6] was the first to prevent and treat successfully meningeal leukemia with intrathecal application of

[1] With support of Deutsche Krebshilfe.

MTX as published in 1959. In the same year *Sullivan* et al. [7] published the administration of higher doses of MTX with simultanous 'rescue' with citrovorum factor, however, the wider use of high-dose MTX with citrovorum factor rescue in order to overcome primary and secondary resistance to the drug goes back to *Djerassi* et al. [8] and their first publication in 1966. *Jaffe* and *Rosen* both applied this new therapeutic regimen to the treatment of osteogenic sarcoma, first published by *Jaffe* [9] in 1972 and by *Rosen* et al. [10] in 1974. In Germany, high-dose MTX has been incorporated since 1977 in 3 consecutive osteosarcoma studies along with other substances such as alkylating agents and anthracyclines.

These studies resulted in an impressive improvement in the prognosis of patients with primary osteosarcoma of the bone. As compared to 20% long-term survival with surgical treatment only between 70 and 80% of all patients can now be cured of osteogenic sarcoma [11, 12].

Chemistry

MTX or amethopterin is 4-amino-4-desoxy-methylpteroylglutamic acid (fig. 1). Its molecular structure differs from folic acid in that folic acid has a hydroxyl group in place of the 4-amino group on the pteridine ring and there is no methyl group at the N^{10} position. MTX differs from aminopterin in that the latter is not methylated at the N^{10} position. Hence, the active site of the molecule appears to involve the pteridine ring portion. The molecular weight is 454. MTX is a bicarboxylic weak acid with a pK_as in the range of 4.8–5.5. As predicted by these physicochemical characteristics, MTX is essentially ionized and lipid insoluble at a physiological pH. Solubility in human urine is directly proportional to pH. The drug will precipitate in acid urine when its concentration exceeds 2×10^{-3} mol/l, whereas its solubility at pH 7 is tenfold greater. Since high-dose therapy does not infrequently produce urine drug concentrations in this range, the drug may precipitate within the kidney and cause renal damage. Alkalinization of the urine prevents precipitation and the associated nephropathy. Alkalinization of the urine also ionizes the drug within the tubule, thereby retarding its reabsorption and promoting its net renal clearance [13–16].

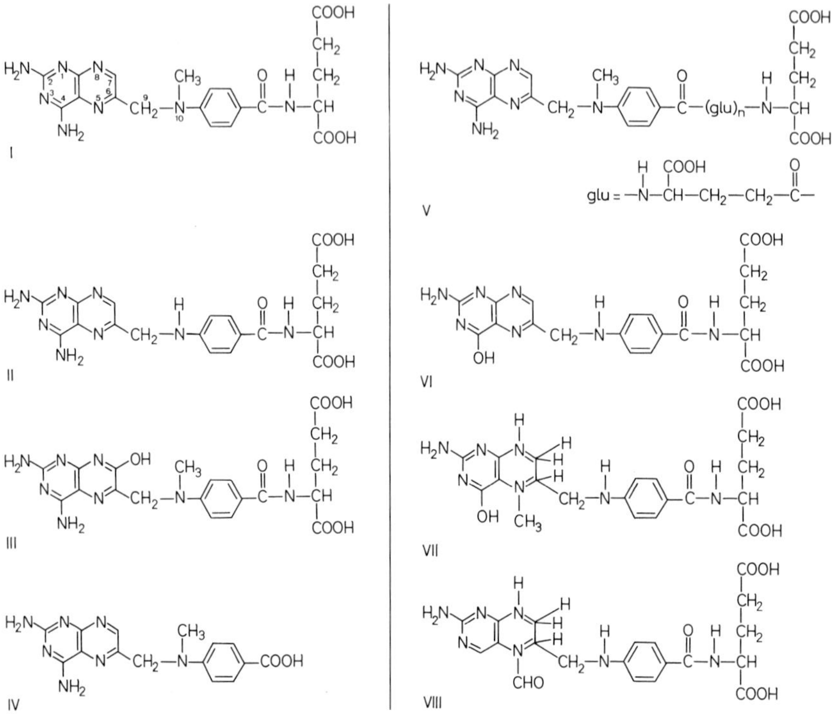

Fig. 1. Chemical formulae of folic acid, antifolates and some derivatives. I = Methotrexate; II = aminopterine; III = 7OH-methotrexate; IV = 4-desoxy-4-amino-N^{10}-methyl pteroic acid (= DAMPA, APA); V = methotrexate polyglutamates; VI = folic acid; VII = tetrahydrofolic acid; VIII = citrovorum factor = 5-formyl-tetrahydrofolic acid.

Mechanism of Action

MTX inhibits dihydrofolate reductase, the enzyme responsible for converting folic acid to reduced folate co-factors (fig. 2). Reduced folates are necessary for the metabolic transfer of one-carbon units in a variety of biochemical reactions. Those reactions which are of special importance in cellular proliferation are the biosynthesis of thymidylic acid, the nucleotide specific to DNA, and the biosynthesis of inosinic

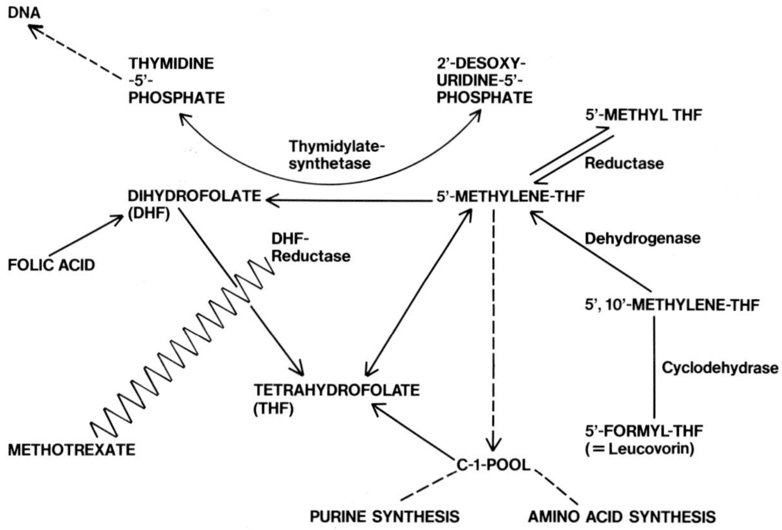

Fig. 2. Folate metabolism. Mechanism of action of MTX. MTX inhibits the intracellular formation of tetrahydrofolate and causes folate depletion. The action of MTX can be bypassed by citrovorum factor.

acid, the precursor of purines necessary for both DNA and RNA synthesis. In human synthesis, MTX appears to inhibit DNA synthesis to a greater extent than RNA synthesis, suggesting that inhibition of thymidylate synthesis is the most important mechanism of MTX cytotoxicity. The drug is highly cell-cycle dependent, acting primarily during DNA synthesis (S phase) [13, 16]. As a result, those tissues undergoing rapid cellular turnover with a high fraction of the cells in cycle are the most susceptible to the drug's cytocidal effects. Low-dose MTX arrests leukemic myeloblasts in the S phase for about 20 h with little effect on cells in G_1, G_2 or M phases of the cell cycle [17]. At higher doses, the arrest is prolonged and the entry of cells from G_1 into the S phase is slowed.

Reduced folates are also required as co-factors for one-carbon unit donation in the conversion of glycine to serine, methionine to homocysteine and histidine to glutamate [13]. By preventing those amino acid interconversions MTX may also interfere with protein synthesis. This effect may be the mechanism by which high-dose MTX arrests cells in the G_1 phase. Cells which are not killed during the S phase at the time of MTX exposure may be protected from fatal injury by their arrest in an earlier phase of the cell cycle and prevention of their progression through the drug-sensitive S phase. According to this hypothesis, high-dose therapy would have a self-limiting effect. On the other hand, cells deprived of protein synthesis for a prolonged interval may be irreversibly altered and eventually succumb as well.

There is also evidence that reduced folates as one-carbon donators are necessary for normal synthesis and metabolism of neurotransmitters in the central nervous system [18]. The implication here is that depletion of reduced folates by MTX may lead to neurotoxic symptoms, particularly when the drug is injected directly into the cerebrospinal fluid. A related important development was the demonstration that intracellular MTX levels in excess of those required to saturate the tight binding sites are necessary for maximal suppression of DNA synthesis. Higher free drug levels are necessary for inhibition of RNA synthesis and ever higher levels for cessation of protein synthesis. This observation confirms the need to maintain the extracellular MTX concentration above a critical level for maximum effect. It reverses the earlier concept which held that once the reductase was saturated an irreversible inhibition ensued and the drug could be discontinued [13, 16].

The action of MTX can be bypassed by the administration of citrovorum factor (5-formyl-tetrahydrofolate, leucovorin, folinic acid). This is converted to 5-methyl-tetrahydrofolate, which is the predominant reduced folate in the body. This metabolite is then converted to tetra-hydrofolate, or to 5,10-methylene-tetrahydrofolate, which enters the reduced folate cycle distal to the MTX-induced enzymatic block (fig. 2). The ultimate effect is the repletion of intracellular reduced folates and resumption of tetrahydrofolate-dependent macromolecular synthesis. Whether citrovorum can rescue normal cells more than tumor cells is still uncertain. For some tumors it could be shown that reduced membrane carrier capacity is the mechanism of selectivity [13].

The exact mechanism of action of MTX in the treatment of nonmalignant disorders is imprecisely understood. However, its antiinflammatory effect is confirmed by its ability to delay the reflux of macrophages in skin disorders, decrease the exudation, alter chemical stimuli in vitro and impair the predictive vasoactive response to serotonine, histamine and kallikreins. Also, suppression of T and B cell function has been shown in vitro and in vivo [19–22].

Absorption

In doses of less than 30 mg/m² MTX is nearly completely absorbed from the gastrointestinal tract [23]. With doses of 80 mg/m² or more, absorption from the gastrointestinal tract is protracted and incomplete, leading to plasma levels less than one-tenth that achieved after intravenous administration [24]. These data implicate the presence of a saturable intestinal absorption mechanism, possibly an active transport process with low-capacity characteristics. That low doses of MTX can be administered orally or parenterally on an equimolar basis has been confirmed by clinical trials, in which orally administered twice-weekly antifolate therapy was just as effective and just as toxic as parenteral usage [13].

Distribution

After intravenous administration, MTX is distributed rapidly within a volume of 18% of body weight and then within a space of 76% of body weight [24]. These volumes of distribution approximate those of the extracellular space and total body water, respectively. After intravenous administration, the MTX disappearance from plasma is triphasic. The initial half-life is less than 1 h and represents that of distribution. The second half-life has been reported as approximately 2 h and is that of the renal clearance. The third half-life of approximately 10 h may reflect the enterohepatic circulation of MTX and also redistribution from third spaces such as cerebrospinal fluid, pleural and peritoneal cavities especially when increased in ascites or pleural effusion [25]. Those third spaces may act as reservoirs and prolong the presence of MTX in the plasma compartment.

The mechanisms by which MTX enters normal and neoplastic human cells have not been fully characterized. Current evidence suggests that reduced folates necessary for normal cellular metabolism gain entry to the cell via an energy-coupled, carrier-mediated mechanism in the cell membrane. MTX and other folate analogs appear to be transported by the same mechanism. There appears to be a separate carrier-mediated mechanism for efflux of the drug which is sensitive to vincristine [13].

The 'blood-brain barrier' slows entry of systemically-administered MTX into the central nervous system. Under steady state conditions the CSF plasma ratio is 0.02–0.05. The ratio appears to be relatively independent of plasma drug levels, and hence CSF MTX concentrations in excess of 10^{-5} mol/l can be achieved by maintaining the plasma level above 5×10^{-4} mol/l for at least 12 h [26].

50–70% of the drug is bound to protein principally albumin in human plasma [23, 24]. Binding to proteins in interstitial fluid effusions is variably low ranging from 0 to 17%. Alterations in plasma protein binding affect the amount of free extracellular MTX, which in turn influences the influx of MTX into cells and its rate of clearance by the kidneys.

Metabolism

It was initially thought that MTX was not metabolized to any significant degree in man [24]. With normal renal function, drug clearance from plasma is above 108 ml/min/m^2 and approximately 40% of an intravenously administered dose is excreted unchanged in the urine within 6 h after administration, 90% within 24 h and 95% within 30 h. At very low plasma concentrations, MTX appears to be reabsorbed by the kidney. At higher concentrations the renal clearance of MTX is relatively constant and exceeds that of inulin clearance, indicating that the antifolate is not only filtered but is also actively secreted by renal tubular cells [14].

Fecal excretion after oral administration is proportional to dosage as predicted by the limitations of intestinal absorption, with values of between 4 and 5% of the total dose of 30 mg/m^2 and approximately 30% of a total dose above 80 mg/m^2.

Recently, several metabolites or contaminants have been found in human urine and plasma, particularly in patients receiving high doses of drug [13, 27]. 7-hydroxy MTX has been found in the urine of patients treated with high doses. The hydroxylase capable of this biotransformation has been found in human liver but not in kidney tissue.

The 7-hydroxy metabolite is found in human urine during the terminal phase of MTX disappearance from plasma, when it may account for 7–33% of the drug excreted during this interval. The 7-hydroxy derivative is four times less soluble in human urine than the parent compound suggesting that the metabolite may be contributing to the nephropathy of high-dose MTX therapy. The 7-hydroxy metabolite does not seem to have any antifolate activity.

Other studies suggested that MTX is metabolized during its enterohepatic circulation by bacterial organisms within the intestinal tract. Their carboxypeptidase splits MTX into glutamate and into the 4-amino-4-desoxy-N^{10}-methylpteroyl-acid metabolite also known as APA or DAMPA. It is important to know that this metabolite interferes with most assays for MTX resulting in too high levels [13].

Polyglutamate derivatives of MTX have been found in various tissues [28]. There is some evidence that these polyglutamates bind irreversibly to the dehydrofolate reductase and may be responsible for some long-term toxicity of MTX.

Drug Resistance

True resistance of certain tumors and normal cells to MTX is most likely due to the low growth fraction and excess population that characterizes most resistant cell population [16]. Also, cells with primary resistance appear to have less membrane carrier activity and therefore the intracellular MTX concentration following conventional dosis is insufficient to completely block the dihydrofolate reductase. Secondary resistance is explained in some tumors by increased synthesis of the enzyme dihydrofolate reductase based on gene amplification [29].

Drug Interactions

A variety of interactions between MTX and other drugs has been described in man or in studies of human tumor cells. Renal tubular transport is diminished by salicylate, sulfamethooxazol, sulfonamides, probenicid and by other weak organic acids [16]. A significantly higher incidence of severe leukopenia was shown in patients taking acetylic acid during MTX therapy. Hydrocortisone and cephalosporins decrease the uptake of MTX by human leukemic cells [30]. Oral broad-spectrum nonabsorbable antibiotics decrease the intestinal absorption of MTX. Antibiotics may also interfere with the enterohepatic circulation of MTX by inhibiting bowel flora and suppressing the metabolism of the drug by these bacteria. Vincristine was shown to reduce cellular efflux of MTX [13, 16].

An interaction that has clinical application is that between MTX and asparaginase [16]. Asparaginase inhibits protein syntheses, arrests cells during the G_1 phase, prevents cells from entering the S phase and thus can protect cells from the S phase effects of MTX. When given in combination with MTX and in the correct interval and sequence, asparaginase is able to limit the bone marrow toxicity due to the antifolate. In view of its own antitumor activity against acute lymphoid malignancies, asparaginase will have a substantial theoretical advantage over citrovorum factor in these diseases. Synergistic effects have also been shown between MTX and other purine antagonists such as cytosine arabinoside, 5-fluoro-uracil, and 6-mercaptopurin.

Usage

There are 3 basic treatment regimens in current use [13]:

(1) Conventional low dosage without citrovorum factor rescue. MTX is applied orally, intravenously or intramuscularly at a dose of between 10 and 50 mg/m^2 per week or as a daily dose of 1–3 mg/m^2/day. Those regimens are applied in the treatment of ALL, AML, choriocarcinoma, breast cancer, prevention of graft-versus-host disease following bone marrow transplantation, and also in nonmalignant diseases such as severe psoriasis, dermatomyositis and rheumatoid arthritis.

(2) Intrathecal administration at a dose of 6–15 mg (age dependent) is used to treat or prevent malignant tumor spread, e.g. in ALL, AML or parameningeal rhabdomyosarcoma.

(3) The higher dose regimens that require citrovorum factor rescue are subdivided in an intermediate dose range between 0.5 and 1 g/m^2 and a high dose range above 1 g/m^2.

The high-dose regimens can further be subdivided into those in which the antifolate is administered over a period of 4–6 h and those in which MTX is infused over 24–42 h. The first regimen results in peak

MTX plasma levels of 10^{-2} to 10^{-4} mol/l with an exponentional decay thereafter. The long-term infusion regimens result in fairly constant plasma levels during the infusion with plateau concentrations ranging between 10^{-4} and 10^{-6} mol/l depending on the dose applied. The rationale for the short infusion is based upon the possibility that in transport-resistant tumors intracellular drug concentrations can be maximised by passive diffusion. The rationale for the long-term infusion regimen is to encompass more of the cell cycle of neoplastic cells to delay citrovorum factor rescue and to reach an equilibrium between plasma concentrations and the level of the drug in interstitial and intercellular compartments. This regimen has been used successfully to overcome secondary resistance by increased production of dehydrofolate reductase. Doses applied in high-dose MTX regimens vary between 6 and 12 g/m^2. Doses applied for long-term infusions vary between 0.5 and 30 g/m^2 [13].

Toxicity

The principal acute toxicities of MTX are myelosuppression, gastrointestinal mucositis and hepatitis. Other side effects depend on the type of regimen administered [13, 16].

Long-term toxicities include megaloblastic anemia from the folate deprivation of bone marrow stem cells. The mild degree of alopecia is usually reversible [31].

The side effects visible on the skin include desquamative dermatitis especially on areas exposed to the sun, vasculitis of the hands and feet, and generalized folliculitis, the latter usually restricted to HDMTX treatment.

Long-term low-dose administration has been associated with liver fibrosis, on rare occasions progressing to liver cirrhoses [32, 33]. In severe liver disease there may be signs of coagulopathy, on rare occasions associated with bleeding disorders, again restricted to HDMTX treatment [34]. Hepatic dysfunction is usually related to the regimen applied. Single injections of MTX occasionally result in transient elevations of serum transaminases. Daily low dose administration at a frequency of less than 12 days/month produces a more prolonged transaminasemia but minimal changes in liver histology. Administration of the same dose more frequently than 12 days/month often results in irreversible hepatic dysfunction with evidence of fibrosis and cirrhosis in liver biopsy specimens. Large doses over 1 or 2 days produce much less hepatotoxicity than smaller, sustained doses over several days [35].

Chronic cough may indicate interstitial pneumonitis [13]. Bone pain may be related to MTX-induced osteoporosis following long-term use. The immunosuppression induced by MTX is milder than that of other chemotherapeutic agents. B cell function seems to be more affected than T cell function [13, 36].

A necrotizing encephalopathy may occur months to years after the onset of MTX therapy [18, 37]. The clinical pattern is one of progressive neurologic deterioration beginning incidently and evolving into severe dementia, atoxia, seizure, and coma. In addition to systemic application of MTX cranial irradiation and or intrathecal MTX are associated risk factors. The risk appears to be proportional to the amount of therapy given. *Aur* et al. [38] noted that encephalopathic reactions developed almost exclusively in patients receiving five or more doses of intrathecal MTX and more than 40 mg/m^2 weekly of intravenous MTX [38].

Renal failure may be seen following high-dose MTX systemic therapy [13]. Nephrotoxicity is particularly dangerous because the drug is retained as a result of renal dysfunction and further toxicity may ensue. Prolonged MTX exposure following renal failure is the major cause of mortality from HDMTX treatment and may occur in about 2% of all patients treated [39].

The risk of side effects of MTX treatment [13] depends on the following: (1) A critical minimal extracellular concentration threshold for each target organ must be exceeded before the organ will be damaged. (2) A critical minimal time threshold (duration of exposure) for each tissue or organ system must be exceeded for toxicity to occur [16]. (3) Both, the concentration threshold and the time threshold must be exceeded for toxicity to occur. (4) The severity of toxicity is directly proportional to the duration of MTX exposure beyond the time threshold and relatively less dependent on the extent of MTX elevation above the extracellular concentration threshold. The severity of toxicity is therefore determined more by the extent to which the time threshold is exceeded than by the magnitude of the elevation above the concentration threshold.

These principles must be observed when MTX is administered. Enhanced understanding of the clinical pharmacology of MTX and the application of this knowledge to the individual patient will reduce further the morbidity of this agent. The efficacy of therapy will benefit from an expanded knowledge of the drug's intracellular pharmacokinetics in the target cells. The important and practical consequences of these advances will be a significant improvement in the therapeutic index of MTX therapy for all patients with antifolate-responsive malignant and nonmalignant diseases.

References

1 Leuchtenberger, C.; Lewisohn, R.; Laslo, D.; Leuchtenberger, R.: Folic acid – a tumor growth inhibitor. Proc. Soc. exp. Biol. Med. *55:* 204–205 (1944).

2 Seeger, D.R.; Smitz, J.M., Jr.; Hultquist, M.E.: Antagonist for pteroylglutamic acid. J. Am. chem. Soc. *69:* 2567 (1947).

3 Farber, S.; Diamond, L.K.; Mercer, R.D.; Sylvester, R.F.; Wolff, J.A.: Temporary remission in acute leucemia in children produced by folic acid antagonists, 4-aminopteroyl-glutamic acid (aminopterin). New Engl. J. Med. *238:* 787–793 (1984).

4 Franklin, A.L.; Belt, M.; Stokstadt, E.; Jukes, T.: Biological studies with 4-amino-10-methylpteroylglutamic acid. J. biol. Chem. *177:* 621–629 (1949).

5 Li, M.C.; Hertz, R.; Spencer, D.B.: Effect of methotrexate therapy upon choriocarcinoma and chorioadenoma. Proc. Soc. exp. Biol. Med. *93:* 361–366 (1956).

6 Murphy, M.L.: Leukemia and lymphoma in children. Pediat. Clins N. Am. *6:* 611–638 (1959).

7 Sullivan, R.D.; Miller, E.; Sikes, M.P.: Antimetabilite-metabolite combination cancer chemotherapy. Effects of intraarterial methotrexate – intramuscular citrovorum factor therapy in human cancer. Cancer *12:* 1248–1262 (1959).

8 Djerassi, J.; Abir, E.; Royer, G., Jr.; Treat, C.: Long-term remissions in childhood acute leukemia. Use of infrequent infusions of methotrexate. Supportive role of platelet transfusions and citrovorum factor. Clin. Pediat. *5:* 502–509 (1966).

9 Jaffe, N.: Recent advances in the chemotherapy of metastatic osteogenic sarcoma. Cancer *30:* 1627–1631 (1972).

10 Rosen, G.; Suwansirikul, S.; Kwon, C.; Tan, C.; Wu, S.J.; Beattiue, E.J., Jr.; Murphy, M.L.: High-dose methotrexate with citrovorum factor rescue and adriamycin in childhood osteogenic sarcoma. Cancer *33:* 1151–1163 (1974).

11 Winkler, K.; Beron, G.; Kotz, R.; Salzer-Kuntschik, M.; Beck, J.; Beck, W.; Brandeis, W.; Ebell, W.; Erttmann, R.; Göbel, U.; Havers, W.; Henze, G.; Hinderfeld, L.; Höcker, P.; Jobke, A.; Jürgens, H.; Kabisch, H.; Landbeck, G.; Preusser, P.; Prindull, G.; Ramach, W.; Ritter, J.; Sekerea, J.; Treuner, J.; Wüst, G.: Adjuvant chemotherapy in osteosarcoma. Effect of cis-platinum, BCD, and fibroblast interferon in sequential combination with HD-methotrexate and adriamycin. Preliminary results of the COSS 80 study. Cancer Res. clin. Oncol. *106:* suppl., pp. 1–7 (1983).

12 Winkler, K.; Beron, G.; Kotz, R.; Salzer-Kuntschik, M.; Beck, J.; Beck, W.; Brandeis, W.; Ebell, W.; Erttmann, R.; Göbel, U.; Havers, W.; Henze, G.; Hinderfeld,

L.; Höcker, P.; Jobke, A.; Jürgens, H.; Kabisch, H.; Preusser, P.; Prindull, G.; Ramach, W.; Ritter, J.; Sekera, J.; Treuner, J.; Wüst, G.; Landbeck, G.: Neoadjuvant chemotherapy for osteogenic sarcoma. Results of a cooperative German/Austrian study. J. Clin. Oncol. *2:* 617–624 (1984).

13 Jürgens, H.: Hochdosierte Methotrexatbehandlung (Urban & Schwarzenberg, München 1983).

14 Liegler, D.G.; Henderson, E.S.; Hahn, M.A.; Oliverio, V.T.: The effect of organic acids on renal clearance of methotrexate in man. Clin. Pharmac. Ther. *10:* 849–857 (1969).

15 Stoller, R.G.; Jacobs, S.A.; Drake, J.C.; Lutz, R.J.; Chabner, B.A.: Pharmacokinetics of high-dose methotrexate (NSC-740). Cancer Chemother. Rep. *6:* 19–24 (1975).

16 Bleyer, W.A.: The clinical pharmacology of methotrexate. New applications of an old drug. Cancer *41:* 36–51 (1978).

17 Ernst, P.; Killman, S.: Perturbation of generation cycle of human leukaemic myeloblasts in vivo by methotrexate. Blood *38:* 689–705 (1971).

18 Abelson, H.T.: Methotrexate and central nervous system toxicity. Cancer Treat. Rep. *62:* 1991–2001 (1978).

19 Willkens, R.; Watson, M.: Methotrexate. A perspective of its use in the treatment of rheumatic diseases. J. Lab. clin. Med. *100:* 314–321 (1982).

20 Willkens, R.F.: Reappraisal of the use of methotrexate in rheumatic disease. Am. J. Med. *75:* 19–25 (1983).

21 Steinsson, K.; Weinstein, A.; Korn, J.; Abeles, M.: Low dose methotrexate in rheumatoid arthritis. J. Rheum. *9:* 860–866 (1982).

22 Willkens, R.F.; Williams, H.J.; Ward, J.R.; Egger, M.J.; Reading, J.C.; Clements, P.J.; Cathcart, E.S.; Samuelson, C.O.; Solsky, M.A.; Kaplan, S.B.; Guttadauria, M.; Halla, J.T.; Weinstein, A.: Randomized, double-blind, placebo-controlled trial of low dose pulse methotrexate in psoriatic arthritis. Arthritis Rheum. *27:* 376–381 (1984).

23 Wan, S.H.; Huffman, D.H.; Azarnoff, D.L.; Stephens, R.; Hoogstraten, B.: Effect of route of administration and effusions on methotrexate pharmacokinetics. Cancer Res. *34:* 3487–3491 (1984).

24 Henderson, E.S.; Adamson, R.H.; Oliverio, V.T.: The metabolic fate of tritiated methotrexate. II. Absorption and excretion in man. Cancer Res. *25:* 1018–1024 (1965).

25 Huffman, D.H.; Wan, S.H.; Azarnoff, D.L.; Hoogstraten, B.: Pharmacokinetics of methotrexate. Clin. Pharmac. Ther. *14:* 572–570 (1973).

26 Jacobs, S.A.; Bleyer, W.A.; Chabner, B.A.; Johns, D.G.: Altered plasma pharmacokinetics of methotrexate administered intrathecally. Lancet *1:* 465–466 (1975).

27 Jacobs, S.A.; Stoller, R.G.; Chabner, B.A.; Johns, D.G.: 7-Hydroxymethotrexate as a urinary metabolite in human subjects and rhesus monkeys receiving high-dose methotrexate. J. clin. Invest. *57:* 534–538 (1976).

28 Covey, J.M.: Polyglutamate derivatives of folic acid coenzymes and methotrexate. Life Sci. *26:* 665–678 (1980).

29 Nunberg, J.H.; Kaufman, R.J.; Schimke, R.T.; Urlaub, G.; Chasin, L.A.: Amplified dihydrofolate reductase genes are localized to a homogeneously stained region of a single chromosome in a methotrexate-resistant Chinese hamster ovary cell line. Proc. natn. Acad. Sci. USA *75:* 5553–5556 (1978).

30 Bender, R.A.; Bleyer, W.A.; Frisby, S.A.; Oliverio, V.T.: Alterations of methotrexate uptake in human leukemia cells by other agents. Cancer Res. *35:* 1305–1308 (1975).
31 Delmonte, L.; Jukes, T.H.: Folic acid antagonists in cancer chemotherapy. Pharmac. Ref. *14:* 91–135 (1962).
32 Nyfors, A.; Poulsen, H.: Liver biopsies from psoriatics related to methotrexate therapy. II. Findings before and after methotrexate therapy in 88 patients. A blind study. Acta pathol. microbiol. scand., A, Pathol. *A84:* 262–270 (1976).
33 Nyfors, A.; Poulsen, H.: Morphogenesis of fibrosis and cirrhosis in methotrexate-treated patients with psoriasis. Am. J. surg. Path. *1:* 235–243 (1977).
34 Göbel, U.; Jürgens, H.; Voss, H. von; Wahn, V.; Rosen, G.: Gerinnungsstörungen nach hochdosiertem Methotrexat mit citrovorum factor rescue bei Patienten mit osteogenem Sarkom. Klin. Pädiat. *193:* 94–98 (1981).
35 Nesbit, M.; Krivit, W.; Heyn, R.; Sharp, H.: Acute and chronic effects of methotrexate on hepatic, pulmonary and skeletal systems. Cancer *37:* 1048–1054 (1976).
36 Mitchell, M.S.; Wade, M.E.; DeConti, R.C.; Bertino, J.R.; Calabresi, P.: Immunosuppressive effects of cytosine arabinoside and methotrexate in man. Ann. intern. Med. *70:* 535–547 (1969).
37 Allen, J.C.: The effects of cancer therapy on the nervous system. J. Pediat. *93:* 903–909 (1978).
38 Aur, R.J.A.; Simone, J.V.; Verzosa, M.S.; Hustu, H.O.; Barker, L.F.; Pinkel, D.P.; Rivera, G.; Dahl, G.V.; Wood, A.; Stagner, S.; Mason, C.: Childhood acote lymphocytic leukemia (study VIII). Cancer *42:* 2123–2134 (1978).
39 Von Hoff, D.D.; Penta, J.S.; Helman, L.J.; Slavid, M.: Incidence of drug related deaths secondary to high-dose methotrexate and citrovorum factor administration. Cancer Treat. Rep. *61:* 745–748 (1977).

PD Dr. Herbert Jürgens, Zentrum für Kinderheilkunde der Universität Düsseldorf, Abteilung für Hämatologie und Onkologie, Moorenstrasse 5, D-4000 Düsseldorf (BRD)

Rheumatology, vol. 9, pp. 20–31 (Karger, Basel 1986)

High-Dose Tumor Therapy with Methotrexate and Its Side Effects

Hansjörg Sauer[a,b], *Andreas Schalhorn*[a], *Wolfgang Wilmanns*[a,b]

[a] Medizinische Klinik III (Direktor: Prof. Dr. *W. Wilmanns*) im Klinikum Grosshadern der Ludwig-Maximilians-Universität, München;
[b] Abteilung für Klinische Hämatologie (Leiter: Prof. Dr. *W. Wilmanns*) des Instituts für Hämatologie der Gesellschaft für Strahlen- und Umweltforschung mbH, München, BRD

Within the different groups of substances which are used in the cytostatic treatment of malignant dieseases, the antimetabolite methotrexate (MTX) has a unique position. No other cytostatic drug can be used over such a wide dose-range. The highest applicable doses are up to three logs more than the usual standard dose.

The definition of different dose levels of MTX is summarized in table I. The low-dose treatment is used during longtime therapy, e.g. in the maintenance phase of acute leukemia of in the immunosuppressive treatment of immunological diseases. The normal dose ist integrated in a variety of different polychemotherapy protocols for solid tumors and systemic malignant dieseases, e.g. breast cancer, lung cancer, malignant lymphomas. The medium dose is escalated to about ten times the normal dose. It is used especially in combination with 5-fluorouracil in adenocarcinomas of the gastrointestinal tract and in squamous cell carcinomas of the ear, nose and throat region. A further dose escalation up to 1000 times the normal dose leads to the high-dose MTX treatment.

The medium-dose and high-dose treatment schedules can only be applied when they are used in combination with a specific antidote of MTX. The antidote is a folic acid derivative, citrovorum factor = 5-formyl-tetrahydrofolic acid = folinic acid = Leucovorin®, which bypasses the antimetabolic effect of MTX. This procedure which prevents severe toxicity after the high doses of MTX is the so-called rescue effect. Table II gives the abbreviations which are used to describe this treatment regimen.

Table I. Dose range in MTX treatment schedules

Range	Dose	Dose factor
Low dose	5–20 mg per dose weekly	ca. 0.1
Normal dose	40–80 mg/m² every 2–4 weeks (mostly in combination chemotherapy)	ca. 1
Medium dose	100–1,500 mg/m² weekly or every 2–4 weeks (mostly in combination chemotherapy)	ca. 10[1]
High dose	8–12 (–50) g/m² weekly or every 2–4 weeks (mostly alternating with other chemotherapy regimens)	ca. 100[1] (– 1,000)[1]

[1] Regimens with antidote rescue.

Table II. Abbrevations used for treatment regimen

HDMTX	=	High-dose methotrexate
CFR	=	Citrovorum factor rescue
	=	Folinic acid rescue
	=	Leucovorin® rescue

To understand how MTX works antimetabolically and how its effect is bypassed by citrovorum factor, the folic acid dependent steps in the DNA synthesis must be described in more detail. Figure 1 shows in a very simplified manner some essential steps in this metabolic pathway.

Dihydrofolate reductase, plays a key role in the metabolism of folic acid. This enzymatic step reduces dihydrofolic (FH_2) acid to tetrahydrofolic acid (FH_4). Tetrahydrofolic acid is the metabolically active form of folic acid and generates the pool of the so-called activated one-carbon (C_1) units which are responsible for the transfer of C_1 units to other molecules in the biosynthesis on purines (adenine, guanine) and the DNA-specific pyrimidine thymidine. MTX specifically inhibits the dihydrofolate reductase activity. This leads to a decrease in the pool of activated one-carbon units which results in a lack of synthesis of substrates for the DNA synthesis.

Independently of the action of MTX, sufficient doses of exogenously applied citrovorum factor, which is itself an activated one-car-

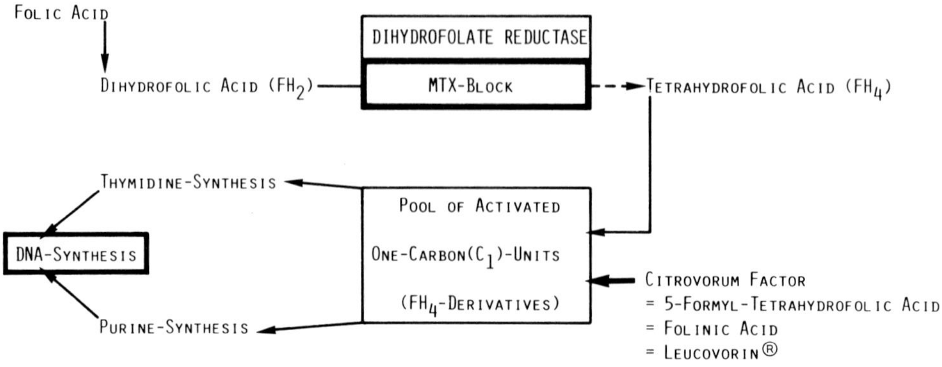

Fig. 1. Tetrahydrofolate-dependent steps in DNA synthesis.

bon unit, can directly restore the pool of the one-carbon units. So, this antidote bypasses the antifolate effect of MTX even if the antimetabolite is still present. Tumors which are resistant to normal doses of MTX have either an increased activity of dihydrofolate reductase (overproduction of enzyme protein due to the amplification of the dihydrofolate reductase coding gene) or have a defective membrane transport system for folate compounds which is used also for the MTX uptake. In both cases very high doses of MTX can overcome these resistance mechanisms and cause a cytostatically effective inhibition of the dihydrofolate reductase activity. As one example, figure 2 describes the principle of the action of high doses of MTX and the citrovorum factor rescue in normal cells as compared to tumor cells with a defective folate transport system.

The normal cells with an active transport capacity accumulate cytotoxic doses of MTX with low doses as well as with high doses. Citrovorum factor (= Leucovorin) given in low doses also enters the cells and can rescue these cells. But, in the transport defective tumor cells, only with high doses of MTX cytotoxic intracellular concentrations can be achieved. With the low doses of Leucovorin, which are administered to rescue the normal cells, no sufficient intracellular concentrations for the rescue of the tumor cells can be established. Thus, selectively tumor cells can be killed by this procedure without any side effects on the normal cell populations which are sufficiently protected by the MTX antidote.

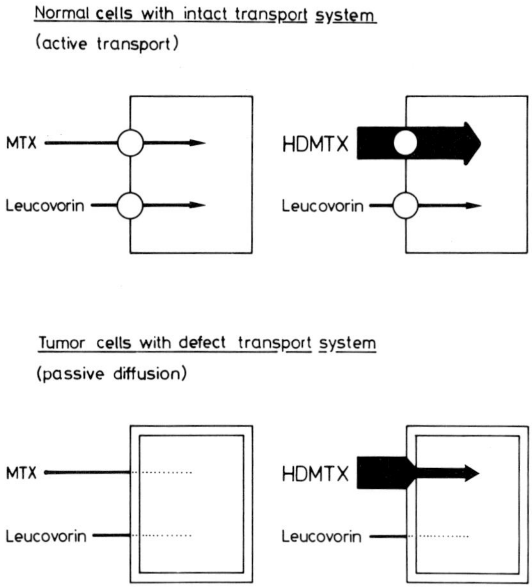

Fig. 2. Principle of high-dose methotrexate (HDMTX) therapy.

Table III shows which tumors can be primarily resistant to normal doses of MTX or develop such a resistance. These tumors respond in different rates to the treatment with HDMTX/CFR.

Figure 3 shows one of many different modes of the clinical application of HDMTX/CFR. It must be assured that the patients have normal renal clearances as well as normal blood counts and a regular liver function. Already before the beginning of the MTX infusion a forced diuresis and urine alkalinization with sodium bicarbonate must be established to protect the kidneys from MTX toxicity and to prevent the crystallization of MTX at high concentrations into acid glomerulum filtrates. After the application of MTX, the antidote, Leucovorin, is given 8 times every 3 h and thereafter 8 times every 6 h. Many modifications of such treatment schedules exist and cannot be described here in more detail. Besides the protection of kidney function, the forced diuresis causes the rapid elimination of the MTX from the body. This elimination must be controlled after each application of HDMTX. Thus, *careful monitoring of serum MTX levels is mandatory when HDMTX with CFR treatment is administered!*

Table III. High-dose MTX folinic acid therapy [from *Frei, E., III,* et al.: Am. J. Med. *63:* 370–376 (1980), modified]

	Effect
Gestational trophoblastic tumors	+ + +
Bone sarcoma	
Adjuvant	+ + +
Palliative	+ +
Head and neck carcinoma	+ +
Ovarian carcinoma	+ +
Lung carcinoma	
Small cell	+
Non-small cell	(+)
Non-Hodgkin's lymphomas	+
Acute leukemia	+
Breast carcinoma	+
Melanoma	(+)
Soft tissue sarcoma	(+)
Gastrointestinal carcinoma	(+)
Genitourinary carcinoma	(+)
Other solid tumors	(+)

+ + + > 50% response; + + < 50% response; + < 25% response; (+) = sporadic response.

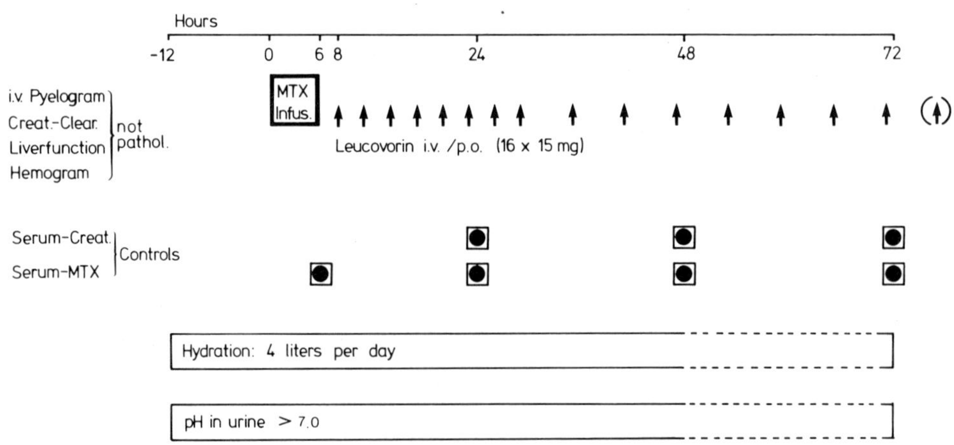

Fig. 3. Therapeutic regimen for high-dose methotrexate (HDMTX) therapy.

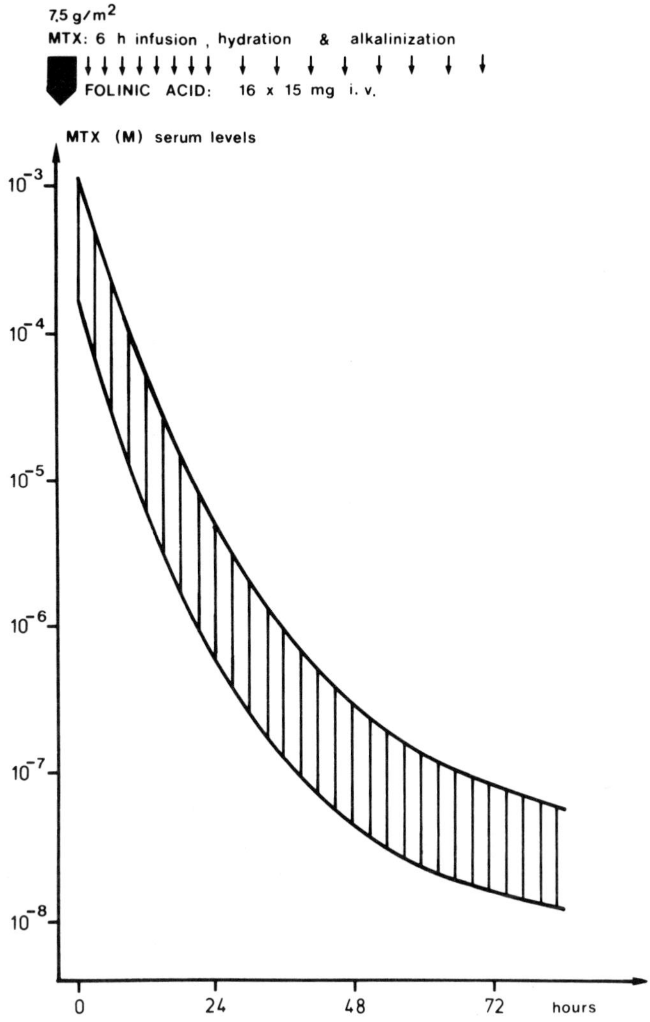

Fig. 4. Normal MTX eliminaton.

Figure 4 shows the normal range of MTX elimination from the blood serum after HDMTX treatment. If the serum levels at the respective time points are within these limits, the routinely administered doses of Leucovorin are sufficient to rescue the normal rapidly proliferating tissues such as bone marrow and mucosae. But, as shown in figure 5, in some cases the MTX elimination is slower.

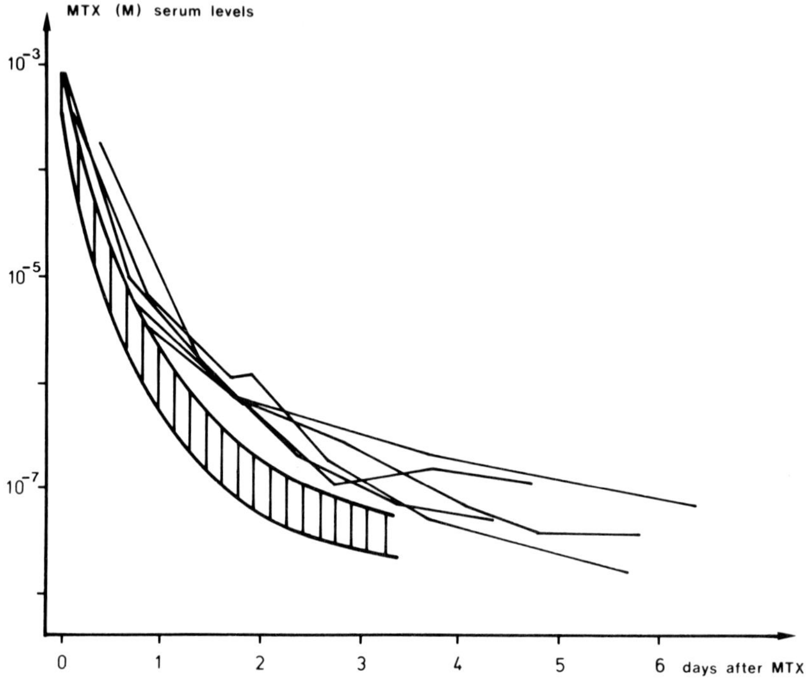

Fig. 5. Retarded MTX elimination.

These patients need an increased dose of Leucovorin for the effective protection of normal tissues. According to biochemical investigations on cell cultures and on human bone marrow cells under HDMTX/CFR treatment, we developed the following formula to calculate the individual doses for the high-dose Leucovorin rescue in the case of retarded MTX elimination:

$$\text{Leucovorin (mg)} = 10 \cdot \text{MTX (mg/liter serum)} \cdot 0.76 \cdot \text{body weight (kg)}$$

The calculated dose is administered every 6 h until the next MTX serum level determination (usually 4 doses, then new calculation with new MTX serum level).

Figure 6 demonstrates the application of this formula in a patient with slow MTX elimination. He received Leucovorin doses which were up to 21 times higher than the routinely administered dose. In the following days no toxicity was observed in this patient.

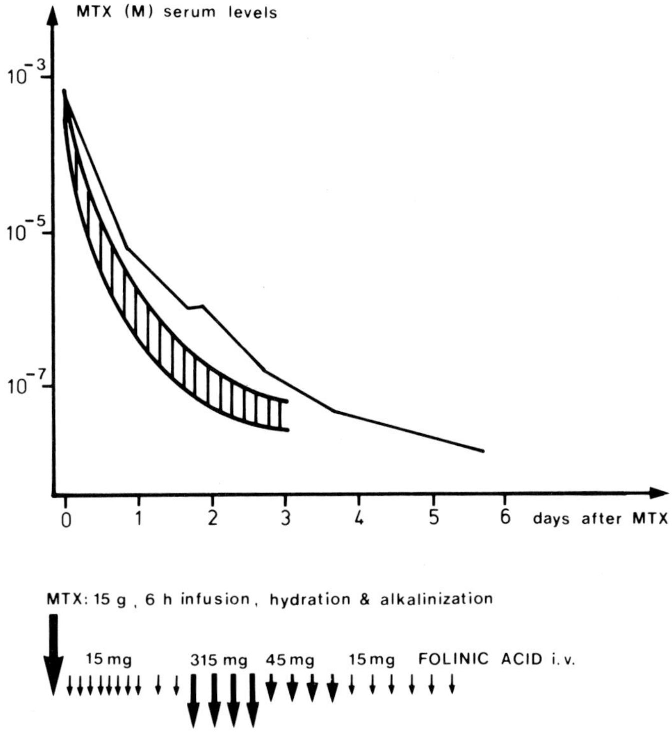

Fig. 6. High-dose folinic acid rescue.

It is important to know that there are several circumstances under which the MTX clearance from the body can be decreased. These are summarized in table IV. The most important are impaired renal function, the redistribution from third spaces and the interaction with other drugs which is described in more detail at this symposium by Dr. *Jürgens.* A special problem is the unknown mechanisms. It may happen that a patient eliminates MTX slowly, even if all organ functions are regular and even if he has eliminated with a normal rate at previous HDMTX/CFR treatments. This fact underlines the urgent need for serum level assays during each HDMTX/CFR treatment. If MTX ist eliminated slowly and if the normal cells are not rescued sufficiently by Leucovorin, severe and potentially lethal complications can occur, as listed in table V.

Table IV. MTX whole-body clearance

Decreased	Increased
Impaired renal function	Forced diuresis
Intestinal obstruction	Rapid intestinal passage
Third spaces	Biliary obstruction
Pleural effusions	Severe vomiting
Ascites	Oral application of charcoal
Central nervous system	and/or other absorbents
Interactions with other drugs	
Salicylates	
Probenecid	
Cephalosporines	
Others	
Unknown mechanisms	

Table V. Severe complications of HDMTX/CFR treatment

Bone marrow depression
Granulocytopenia (acute)[1]
Thrombocytopenia (acute)[1]
Anemia (after chronic application)
Mucositis (acute)[1]
Skin rash (acute)
Pneumonitis (acute)[1]
Liver dysfunction (chronic, fibrosis)
Leukoencephalopathy (chronic, in combination with cranial irradiation)[1]

[1]Potentially lethal.

The next figures show two examples of toxic reactions after MTX applications.

Figure 7 shows the course of a patient who was treated in an external department with 450 mg of MTX in spite of impaired renal function with a serum creatinine of 3.1 mg/dl. MTX serum levels were not measured and no increased Leucovorin rescue was performed. The patient developed severe toxicity with bone marrow depression, sepsis and mucositis. Then the patient was admitted to our hospital. As it can be seen from the course of the granulocyte and platelet counts, the already manifest toxicity cannot be reversed by Leucovorin. The patient needed granulocyte and thrombocyte transfusions. But, after the

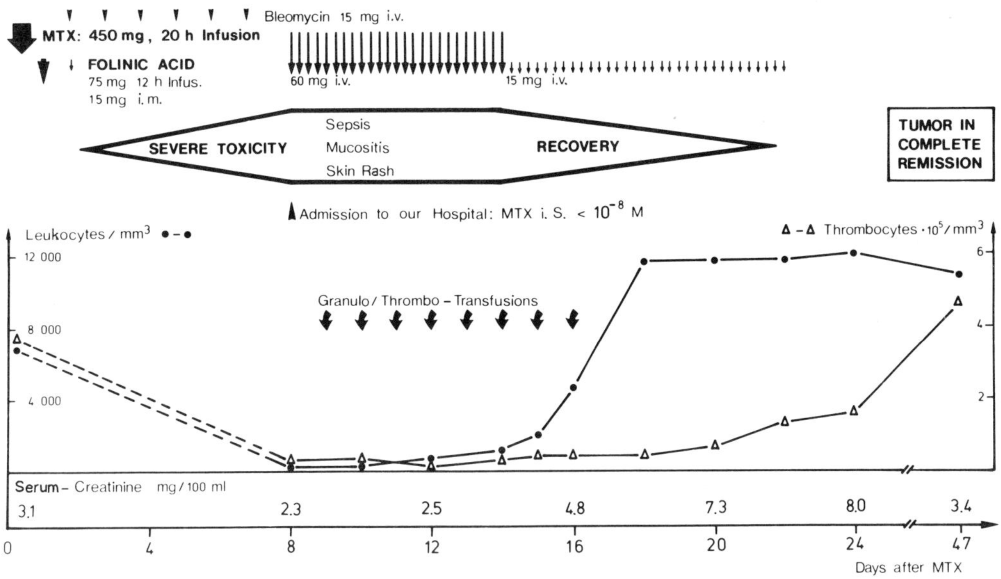

Fig. 7. MTX intoxication in renal insufficiency: 'Bitter'-treatment (patient J. R., male, 48 years). Squamous cell epipharynx carcinoma.

recovery from this very intensive MTX treatment, he had achieved complete remission of his squamous cell carcinoma in the epipharynx.

The second patient, who developed MTX-induced toxicity was treated in our clinic with three intrathecal doses of 15 mg MTX each. She had a meningiosis carcinomatosa and brain metastases from a breast cancer.

After the third injection she developed severe bone marrow suppression and mucositis, as shown in figure 8.

Because it was not a high-dose MTX treatment, the serum levels were not tested prospectively. But when toxicity became evident these assays were performed in frozen serum specimens. It became evident that the patient had a redistribution from her third space in the central nervous system. A disruption of the physiological blood-brain barrier might be responsible for this increased passage of the drug into the blood. Figure 9 shows the polonged exposure to MTX for 2–3 days after each intrathecal injection. The shaded area represents the normal elimination of MTX from the blood after the intrathecal application of the drug.

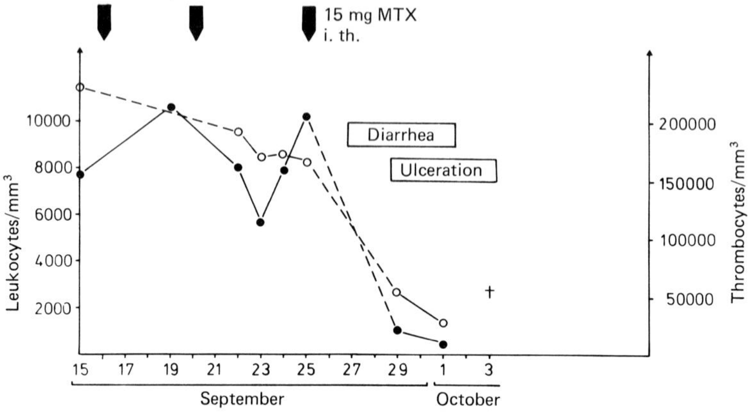

Fig. 8. Patient S. J. Meningiosis carcinomatosa. ● = Leukocytes; ○ = thrombocytes.

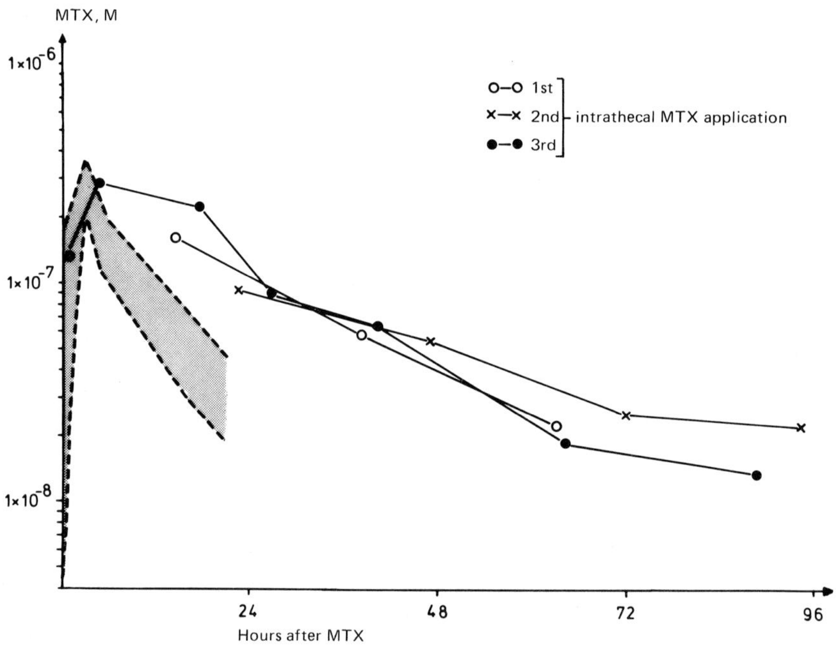

Fig. 9. Patient S. J. MTX serum levels.

Table VI. MTX detoxification

MTX serum level adapted treatment with the specific antidote Citrovorum factor (Leucovorin)
Forced diuresis (6 liters/m^2 daily)
Urine alkalinization (pH over 7.0)
Drainage of third spaces: pleural effusions, ascites
Ventriculolumbar perfusion (if MTX was given intrathecally)
Hemodialysis, charcoal perfusion
To prevent severe bone marrow depression and mucositis, any detoxification measures must begin within at least 48 h after the beginning of the MTX treatment.

Table VII. Summary

HDMTX treatment with CFR is an effective therapeutic regimen for some malignant diseases
Subtile patient selection is critical and MTX serum level monitoring is mandatory
Accidentally HDMTX/CFR can cause potentially lethal complications
Therefore, HDMTX/CFR treatment should be applied only in specialized tumor centers
In the hands of experienced oncologists HDMTX/CFR is almost nontoxic

Sometimes it might happen that by accident, or even in a suicidal intention, high doses of MTX are taken by a patient. In these cases detoxification measures, as summarized in table VI, can prevent severe toxicity.

The detoxification combines the two main principles that are used in HDMTX/CFR treatment: the elimination of the drug and the use of its specific antidote in optimal concentrations. But, if MTX has already caused cell kill in bone marrow and mucosae, a rescue cannot be established by any method. The patient must be kept under intensive supportive care with antibiotics and substitution of blood components until the tissues regenerate from suviving stem cells.

In summary, table VII gives some important statements that should be followed when patients are treated with HDMTX/CFR.

Prof. Dr. med. Hansjörg Sauer, Medizinische Klinik III, Klinikum Grosshadern,
Ludwig-Maximilians-Universität München, Marchioninistrasse 15,
D-8000 München 70 (BRD)

Rheumatology, vol. 9, pp. 32–34 (Karger, Basel 1986)

Discussion 1

Willkens: I'd like to ask both Dr. *Jürgens* and Dr. *Sauer* how they evaluate the evidence on the selective toxicity of the drug for neoplastic versus normal cells. How convincing is the experimental data in their mind? Are neoplastic cells more selectively destroyed than normal cells?

Jürgens: The most convincing evidence is tumor response. Experimental data, as published by *Frey* in 1975 where it was shown in tumor cell lines, that DNA synthesis is resumed in normal bone marrow stem cells when citrovorum factor is added, however, the DNA synthesis in tumor cell lines remained arrested for approximately 9 days. In primary previously untreated osteosarcoma of bone the primary tumor shows a more than 90% destruction with high-dose methotrexate treatment only in approximately 40% of patients, as published by *Rosen.*

Sauer: In addition to the statement of Dr. *Jürgens* I would like to add a case report to demonstrate the effect of high-dose methotrexate with citrovorum factor rescue. This patient with osteogenic sarcoma metastatic to both lungs was treated with high-dose methotrexate only, with citrovorum factor rescue and received no further chemotherapy. High-dose methotrexate produced a remarkable and impressive tumor reduction in this patient and during this treatment, the patient had no evidence of bone marrow depression or stomatitis. This example gives evidence for the selective effect of high-dose methotrexate with citrovorum factor rescue on tumor tissue, with deminished side effects on proliferating normal tissues.

Willkens: To both of the speakers: what I'm interested in is if administering folic acid to people who are taking low-dose methotrexate protects them from side effects. Has the serum folate status of patients, prior to receiving methotrexate in these higher doses for cancer, anything to do with the kind of toxicity, the amount of toxicity, the seriousness of toxicity, or not. *Bextino* seems to think so, and I wonder whether you do. That is, people with subclinical folate deficiency before they get the drug.

Sauer: There is, I think, good evidence that patiens with subclinical folate deficiency react with sustained bone marrow suppression if you give even low doses of methotrexate. And it may be that there are some patients with a defect in the metabolism of folic acid, who don't have the normal activity of the dehydrofolic reductase. And if these patients receive methotrexate, the side effects may be very severe or even lethal.

Willkens: Is it your contention then that perhaps patients ought to be repleted before you start that sort of therapy? Do you think that you should do a serum folate level on patients before you start, and if they are low, give them some folinic acid or something of that sort?

Sauer: I don't think that you need that, because defects in folate metabolism are very rare, the incidence maybe being approximately 1/1 million.

Jürgens: A good example of what has just been mentioned is leukoencephalopathy. It has been shown that leukoencephalopathy only appears following repeated exposure to methotrexate when the brain has no time to recover from folate deprivation and to replete its folate stores. The cases of encephalopathy following high-dose methotrexate treatments all have had 6 of 7 repeat doses of weekly high-dose methotrexate. We have determined the serum and SCF folate levels after various high-dose methotrexate treatments, and we have seen that MTX, as administered routinely, does not enter the CSF to a great extent. At 72 h following high-dose methotrexate treatment systemically always higher folate than methotrexate levels are detected, indicating effective rescue. In the CSF at 72 h there is still more methotrexate than folates. Of course, there is less dividing tissue in the CNS, but there are other side effects, like inhibited neural transmitter or nerve sheaths synthesis, that may result in toxicity.

As for the question of folate repletion prior to initiating high-dose methotrexate treatment, I would argue that the best clinical evidence of restored folates is a moral peripheral blood count, indicating normal bone marrow function and a normal gastrointestinal mucosa. It is quite easy to measure folates, but it is very difficult to measure the essential compounds like citrovorum factor or the activated form 5-methyl-tetrahydrofolic acid. Since only very few laboratories are able to determine those compounds, essaying these substances cannot be made mandatory prior to methotrexate treatment.

Stroehmann: Dr. *Jürgens,* I miss the discussion of carcinogenesis in your paper. Did you avoid it because you did not have enough data about this point?

Jürgens: I didn't mention the carcinogenesis of methotrexate because there is no carcinogenesis.

Stroehmann: Is this also true for lymphoid tumors?

Jürgens: Reviewing the literature of treatment-induced tumors, there is a risk of second malignancies in patients receiving chemotherapy and this risk is correlated with the cumulative dosis of chemotherapeutic agents administered. But the agents that are associated with secondary tumors are alkylating agents, procarbazine and all other agents, that lead to DNA breakage. It is difficult to describe a patient population with malignant tumors that (a) survive the malignant tumor long enough to develop a second malignancy, and (b) have only received antimetabolites. But reviewing the literature on methotrexate, to my knowledge, the risk of methotrexate-associated second neoplasms is not significant.

Stroehmann: You know that, for example, by using the antimetabolite azathioprine, you do have a risk in introducing lymphoma. This is not an alkylating agent but has a different mechanism of action. Could it not be that there is just not enough experience with long-term low-dose methotrexate treatment in order to assess the risk of second malignancies?

Sauer: Maybe. This is very interesting since azathioprine is a drug used by all of you. Azathioprine is not only an antimetabolite, that must be emphazised, there is an imidacole part of its molecule and this is a molecule that reduces the repair capacity of the DNA strand and leads to breakage. And I think that the imidacole part of the molecule is responsible for the carcinogenetic potential of azathioprine and not the 6-mercaptopurine part of the molecule.

Schmidt: I have a very simple pharmacological question to Dr. *Jürgens.* You have mentioned the interactions between salicylates and MTX. Is there any evidence that we should expect similar interactions between MTX and other nonsteroidal drugs, and if this is the case, should we strictly avoid such combinations or should we adapt the dosage of MTX?

Jürgens: At conventional doses the tubular secretion of methotrexate does not play a major role in the renal elimination of the drug. For this reason, one should not anticipate a high degree of drug interaction between aspirin and other weak acids competing with methotrexate for the active membrane carrier of tubular secretion. But my statement may have to be corrected by the experience of other participants of this conference who are using low-dose methotrexate in combination with other nonsteroid medications.

Furst: Are there any comments from the audience on that point?

Fenner: I think it is a question of the dose. Normally, most new anti-inflammatory drugs are used in a much lower dose than aspirin. And then, the interaction mechanisms with the secretion doesn't play any role. I think, it's a good explanation that most of the new drugs do not show any risk of interaction with MTX, except phenoprofene which is dosed up to 2.5 g, but it's not used so widely.

Rau: So, Dr. *Jürgens,* your statement is that drug interactions with MTX can be expected only in the high-dose tumor therapy and not in low-dose MTX treatment. I have a question regarding side effects: Did you personally see lung complications or chronic liver disease in liver biopsy or at necropsy in patiens with high-dose MTX therapy and, in addition, signs of toxicity?

Jürgens: In our patients, on maintenance treatment for acute lymphoblastic leukemia, that received weekly low-dose methotrexate for a period of 18 months and in patients receiving other regimens of methotrexate, either high-dose methotrexate 12 g/m^2 or intermediate dose methotrexate of 0.5 g/m^2, we see elevation of the transaminases following methotrexate, usually ranging between 100 and 200 IU/ml, sometimes 500 IU/ml with a peak at 24 h after the initiation of treatment. And there is usually a complete recovery within 1 week, before the patient is due for the next treatment. In patients on repeated weekly exposure of low-dose methotrexate, the transaminases are elevated usually in the range between 40 and 60 IU/ml during the treatment period and this elevation then resolves within 3 months after the discontinuation of treatment. Since it does resolve in the great majority of patients, we did not obtain histologic proof in patients treated in our institution that there is no long-lasting liver disease resulting from methotrexate treatment. However, we do not have any clinical evidence of disturbed liver function.

As for the methotrexate pulmonary toxicity apart from chronic cough, seen in patients on lowdoses of methotrexate, we occasionally have observed a pneumothorax complicating high-dose methotrexate treatment. It remains unclear whether the pneumothorax results from direct toxic effect of methotrexate, or from dissolving subclinical pleuralmetastases in response to methotrexate.

Furst: Dr. *Sauer,* do you have anything to add before we go on? Another comment?

Sauer: I might add that we see in postmortem liver specimens from these patients a fatty disposition of the liver, but no fibrosis an no cirrhosis. But these are mostly patients who are treated with methotrexate for 6 months or up to 2 or a maximum of 3 years. And maybe I should add that we had 1 patient just a few weeks ago with severe pulmonary toxicity along with other methotrexate side effects. I can't prove that methotrexate caused these two phenomena, but it occurred at the same time when the methotrexate was applied to the patient.

Rheumatology, vol. 9, pp. 35–41 (Karger, Basel 1986)

Pharmacokinetics of Very Low Dose Methotrexate

Daniel E. Furst

Division of Rheumatology, University of Iowa Hospitals, Iowa City, Iowa, USA

Introduction

Methotrexate is an effective drug in rheumatoid arthritis, even when used in very low weekly doses (5–25 mg weekly). On the other hand, it has significant toxicity, such as stomatitis, cirrhosis, bone marrow suppression, lung fibrosis, etc. Little data exists on *how* low a dose is optimal nor how to tailor doses to individual patient needs; for those purposes more data on methotrexate clinical pharmacokinetics is needed.

Absorption

While conventional wisdom says that the bioavailability of methotrexate in doses less than 30 mg/m^2 is complete, data by *Balis* et al. [13] indicate some variability. In 15 children with acute lymphocytic leukemia, in whom methotrexate concentrations where followed for only 6 h, bioavailability ranged from 0.95 to 0.23 as dose rose from 6 to 28.1 mg/m^2. Even within a single patient bioavailability dropped from 0.51 to 0.23 as the dose went from 10.3 to 20.5 mg/m^2. While the differences in mean bioavailability appear quite consistent, there was significant interindividual variability (fig. 1).

In preliminary studies of patients with rheumatoid arthritis given 10 mg/m^2 i.v. and p.o. methotrexate, we found a mean bioavailability of approximately 0.5 [*Furst,* unpubl. data].

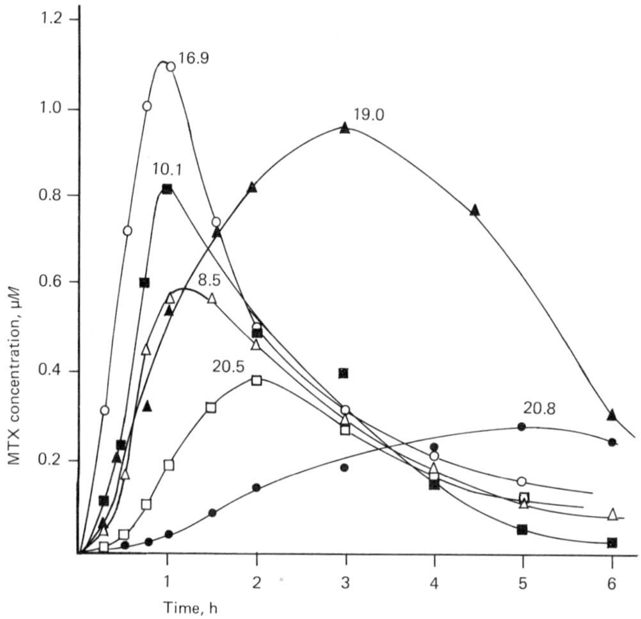

Fig. 1. Serum methotrexate concentrations in 5 patients given differing oral doses of drug. The figure demonstrates the variability of methotrexate kinetics. Number with each curve is the dose in mg/m^2. By permission from *Balis* et al. [1].

Edelman et al. [2] compared intravenous and intramuscular 10 mg methotrexate doses. Their study showed no major differences between these two forms of therapy, but concentrations were only followed for 24 h and no appropriate kinetic comparisons (such as bioavailability) were made. The latter preliminary study is suggestive but certainly not definitive.

Disposition

In 22 patients with malignancy, 30 mg/m^2 intravenous methotrexate disappeared from plasma with 3 half-lives, the terminal one of which was 26.99 ± 4.44 h (SD) [3]. The initial half-life was 0.75 ± 0.11 h and the second half-life was 3.49 ± 0.55 h. Using radioactive counting of tritium labeled methotrexate, the authors found that 20% (range

14–27%) of the radioactivity in urine derived from a metabolite. Thus, this data may be somewhat nonspecific. On the other hand our preliminary data, using a specific radioimmune assay, reveals a mean terminal serum half-life of 5.7 h [*Furst,* unpubl. data].

An enterohepatic recirculation is suggested by data using the *D*-isomer of methotrexate as a reference marker [4]. Assuming that *D*-methotrexate is not absorbed and that no optical isomeratization of *D*- to *L*-methotrexate occurs, Hendel and Brodthagen calculate that the median ratio of biliary to renal excretion of *D*-methotrexate is 0.94 (range 0.41–1.50). Direct measurement of *L*-methotrexate excreted into a post-operative bile duct 'T' tube in one patient found 20% of a 50-mg dose excreted in the bile [5]. These data indicate active biliary excretion and significant enterohepatic recirculation in humans.

Biliary excretion can, to some extent, compensate for decreased renal methotrexate excretion. Measuring tritium-labeled methotrexate, *Steinberg* et al. [6] note an 18% increase in biliary methotrexate excretion when rats are made anephric. Thus, there are compensatory mechanisms by which methotrexate may be excreted if renal clearance decreases. This may be particularly important if salicylates are used concomitantly with methotrexate where interference with renal MTX clearance may occur.

Cellular uptake of methotrexate into perfused isolated rat livers is rapid and 96% complete after 5 h. The maximum velocity of uptake is 11.1 m*M*/h/g liver, with a K_M of 1.3 m*M* [7]. Once in the cell, methotrexate undergoes significant metabolism. In humans given 2–3 g i.v. methotrexate, 7–10% may appear as the 7-hydroxy derivative [8]. Even when given as repeated 0.7–1.2 mg/kg doses in 2 patients, 7-hydroxy methotrexate appeared at concentrations 1–2 times that of the parent compound, after 12–24 h [9]. However, these concentrations were still extremely low and less than 0.5 n*M*/ml. *Johns and Loo* [10] note that 7-OH-MTX *does* inhibit dihydrofolate reductase just as methotrexate does, but its inhibitory constant is 6×10^{-9} *M* in rabbit liver cells (compared to 3×10^{-11} *M* for methotrexate); therefore, it is a very poor inhibitor of dihydrofolate reductase if its concentration is not extremely high.

Methotrexate's concentrations in rat tissue remain constant over a week (e.g. approximately 12 pmol/g in brain, 415 pmol/g in kidney, and 800 pmol/g in liver), a fact explained by methotrexate's polyglutamation [11, 12]. In both rat liver and cultured human breast cancer cells methotrexate was successively polyglutamated [11, 12]. For example, methotrexate polyglutamates account for none of the intracellular methotrexate in breast cancer cells at 1 h but account for 54.7% of the intracellular methotrexate by 24 h [12]. Further, methotrexate polyglutamates affect cell function, for they inhibit tritiated deoxyuridine incorporation (an assay of thymidilate synthesis). Thirty to thirty-five percent inhibition of deoxyuridine incorporation persisted for 24 h after the cells were removed from all methotrexate, thus showing that retained, intracellular, polyglutamated, methotrexate *does* inhibit thymidine synthesis. While prolonged functional inhibition was noted in one cultured breast cancer cells, no prolonged inhibition occurred in 1 of 2 other human breast cancer cell lines tested, thus showing that not all cells polyglutamate methotrexate equally, or that polyglutamated methotrexate is not equally effective in all cell lines.

7-Hydroxy methotrexate is not only polyglutamated in mouse Ehrlich ascites tumor cells in vitro, but it also inhibits dihydrofolate reductase activity [13]. The concentration of 7-hydroxy methotrexate polyglutamates was high and the reaction occurred in vitro in mouse tumor cells, but this elegent study further indicates the need to consider 7-hydroxy methotrexate metabolism in human experiments.

Ten milligrams methotrexate injected intraperitoneally into mice rapidly affected cellular function as measured by tritium incorporation into cellular DNA [14]. DNA synthesis by mouse duodenal cells decreased by 90% within 1 h of the dose and required 24 h to return to normal. A similar rapid decrease in DNA synthesis by bone marrow cells occurred, but DNA function returned to normal by 10 h. If, as is likely, one can extrapolate qualitatively from mice to humans, this experiment indicates that intermittent methotrexate doses affect the bone marrow less than the GI tract and clinical experience indeed support this contention.

Elimination

Biliary excretion of methotrexate has been documented above but renal methotrexate excretion required more consideration.

After 30 mg/m^2 methotrexate in 30 cancer patients, 79.8% of methotrexate or its labeled metabolites are recovered in the urine in 96 h. In contrast, only 1.3% of the radioactivity appears in the feces, leaving

19% unrecovered [3]. However, great variability is noted, for urinary recoveries ranged from 49.3 to 105.6% while fecal excretion accounted for 0 to 6.9% of the dose. *Henderson* et al. [15] documented increased fecal methotrexate excretion after high dose (10 mg/kg). At 0.1–1.0 mg/kg urinary recovery averaged 83% while stool recovery was 4.4% [15]. After 10 mg/kg oral methotrexate in 1 patient, fecal recovery was 39% whereas it was 3% after an equal i.v. dose. The autors postulated that some of the drug is metabolized by gastrointestinal bacteria so that the label was removed and not absorbed. They therefore repeat the study after giving the patient neomycin to clear bowel flora; in this case they found *decreased* absorption so that bacterial metabolism was not a viable explanation. At high dose, the methotrexate itself may affect gastrointestinal function and result in at least transient 'malabsorption' of the drug.

Renal methotrexate clearance apparently involves tubular mechanisms, for methotrexate clearance averaged 126% of inulin clearance in 15 patients given 5–15 mg methotrexate [16]. Methotrexate renal clearance in this study was approximately 180 ml/min. In our 20 patients, renal clearance averaged approximately 88 ml/min after a 10 mg/m^2 i.v. dose [*Furst,* unpubl. data].

p-Amino hippuric acid, an organic acid which was to have been used to measure renal plasma flow, decreased methotrexate clearance by 52%. In 2 patients given sodium salicylate to a steady state serum concentration of 200 and 220 mg/l, methotrexate clearance was 35% less than without the salicylates. At the same time, sodium salicylate decreased methotrexate protein binding from 0.7 to 0.5. While inhibition of methotrexate renal clearance could decrease *total* methotrexate clearance, decreased methotrexate protein binding could increase methotrexate's volume of distribution, and methotrexate biliary excretion and hepatic metabolism could nullify these effects (although this is speculative). While salicylates may increase toxicity after high dose methotrexate therapy, their effects on toxicity after low dose methotrexate are unknown [17]. Further detailed studies of this potentially important drug interaction are clearly needed.

Hendel and Nyfors recently showed that an occasional patient may demonstrate saturable methotrexate renal clearance even at doses used to treat psoriasis and rheumatoid arthritis. In 12 psoriatic patients with normal creatinine clearance (median 94 ml/min, range 62–118 ml/min) 1 of 12 patients clearly showed saturation of methotrexate renal clearance. This patient had only been given 12.5 mg methotrexate. Unfortunately, no correlation between creatinine clearance and the saturation of methotrexate renal clearance was attempted. By one method, *Hendel and Nyfors* [18] found renal methotrexate clearance to average 82.5 ml/min while by another method it was 107 ml/min; both of these numbers are low relative to those found by ourselves and *Liegler* et al. [16] (see above).

Using their own data plus data from the literature, *Hendel and Nyfors* postulated that methotrexate was cleard by a renal secretory process utilizing the general organic acid transport mechanism *plus* an active reabsorptive process unaffected by organic acids. According to their data renal methotrexate clearance in humans is constant at concentrations up to about 0.2 μg/ml and then increases to concentrations up to 10 μg/ml. At concentrations higher than this, nephrotoxicity may occur and renal clearance would then again decrease. While maximum methotrexate concentrations may be as high as 5 μg/ml in treating psoriatics with 25 mg methotrexate, steady-state concentrations this high on the dosing regimes used for psoriasis and methotrexate are very unlikely [18].

Conclusions

Methotrexate kinetics appear complicated, with variable absorption, saturable and perhaps time-related changes in distribution and excretion as well as extensive metabolism. In this setting, extensive, careful pharmacokinetic studies at the low doses used in nonmalignant diseases must be done and, further, they must be correlated to clinical efficacy and toxicity. Only in this way can more rational therapy of these diseases with low-dose methotrexate be developed.

References

1 Balis, F.M.; Savitch, J.L.; Bleyer, W.A.: Pharmacokinetics of oral methotrexate in children. Cancer Res. *43:* 2342–2345 (1983).

2 Edelman, J.; Russell, A.S.; Biggs, D.F.; Rothwell, R.S.; Coates, J.: Methotrexate levels, a guide to therapy. Clin. exp. Rheum. *1:* 153–156 (1983).

3 Huffman, D.H.; Wan, S.H.; Azarnoff, D.L.; Hoogstratem, B.: Pharmacokinetics of methotrexate. Clin. Pharmacol. ther. *14:* 572–579 (1973).

4 Hendel, J.; Brodthagen, H.: Enterohepatic cycling of methotrexate estimated by the use of the *D*-isomer as a reference marker. Eur. J. clin. Pharmacol. *26:* 103–107 (1984).

5 Calvert, A.H.; Bondy, P.K.; Harrap, K.P.: Some observations on the human pharmacology of methotrexate. Cancer Treat. Rep. *61:* 1647–1656 (1977).

6 Steinberg, S.E.; Campbell, C.L.; Bleyer, W.A.; Hillman, R.S.: Enterohepatic circulation of methotrexate in rats in vivo. Cancer Res. *42:* 1279–1282 (1982).

7 Strum, W.B.; Liem, H.H.: Hepatic uptake, intracellular protein binding and biliary excretion of amethopterm. Biochem. Pharmac. *26:* 1237–1240 (1977).

8 Jacobs, S.A.; Stoller, R.G.; Chabner, B.A.; Johns, D.G.: 7-OH-methotrexate as a urinary metabolite in human subjects and rhesus monkeys receiving high dose methotrexate. J. clin. Invest. *57:* 534–538 (1976).

9 Chan, K.K.; Nayar, M.S.B.; Cohen, J.L.: Metabolism of methotrexate in man after high and conventional doses. Res. Commun. chem. Pathol. Biochem. *29:* 551–561 (1980).

10 Johns, D.G.; Loo, T.L.: Metabolite of 4-amino-4-deoxy-N^{10}-methylpterolyglutamic acid (methotrexate). J. pharm. Sci. *56:* 356–359 (1967).

11 Krakower, G.R.; Kamen, B.A.: In situ methotrexate polyglutamate formation in rat tissues. J. Pharmac. exp. Ther. *227:* 633–638 (1983).

12 Jolivet, J.; Schilsky, R.L.; Bailey, B.D.; Drake, J.C.; Chabner, B.A.: Synthesis, retention and biological activity of methotrexate polyglutamates in cultured human breast cancer cells. J. clin. Invest. *70:* 351–360 (1982).

13 Fabre, G.; Fabre, I.; Matherly, L.H.; Canoe, J.-P.; Goldman, I.D.: Synthesis and properties of 7-hydroxymethotrexate polyglutamyl derivatives in Ehrlich ascites tumor cells in vitro. J. biol. Chem. *259:* 5066–5072 (1984).

14 Chabner, B.A.; Young, R.C.: Threshold methotrexate concentration for in vivo inhibition of DNA synthesis in normal and tumorous target tissues. J. clin. Invest. *52:* 1804–1811 (1973).

15 Henderson, E.S.; Adamson, R.H.; Oliverio, V.T.: The metabolic fate of tritiated methotrexate. II. Absorption and excretion in man. Cancer Res. *25:* 1018–1024 (1965).

16 Liegler, D.G.; Henderson, E.S.; Hahn, M.A.; Oliverio, V.T.: The effect of organic acids on renal clearance of methotrexate in man. Clin. Pharmacol. Ther. *10:* 849–857 (1969).

17 Mandel, M.A.: The synergistic effect of salicylates on methotrexate toxicity. Plast. reconstruct. Surg. *57:* 733–737 (1976).

18 Hendel, J.; Nyfors, A.: Nonlinear renal elimination kinetics of methotrexate due to saturation of renal tubular reabsorption. Eur. J. clin. Pharmacol. *26:* 121–124 (1984).

D.E. Furst, MD, Division of Rheumatology, Department of Internal Medicine, University of Iowa Hospitals, East 400th, Iowa City, IA 52242 (USA)

Rheumatology, vol. 9, pp. 42–45 (Karger, Basel 1986)

Immunosuppressive Therapy of Rheumatic Diseases[1]

I. Stroehmann, Bonn

The failure to achieve sufficient therapeutic succes in the treatment of rheumatic diseases with so-called remission-inducing drugs led to the use of immunosuppressive agents. Despite the still questionable assumption that rheumatoid arthritis (RA) and equal diseases are induced by immunological processes, immunosuppression was thought to reduce an otherwise increased immune response.

The general features of an autoimmune disease include malfunctions of the cellular and humoral type. There is sufficient experimental evidence to postulate even for 'pure' cellular systems such as T cells and macrophages humoral factors as suppressor or helper factor, interleucins, etc. (fig. 1). Interestingly most of the immunosuppressive agents act versus cellular mechanisms, whereas suppression of humoral immune factors include only plasmapheresis. Action against cellular mechanisms include such diverse manipulations as thoracic drug drainage, immunostimulative drugs (immunosuppression by activation of T suppressor cells) and cytostatic or cytotoxic drugs (table I). Some of these treatments are rather experimental with a high risk of even lethal side effects (e.g. total lymphoid irradiation) whereas others such as cytostatic drugs represent a considerable proportion of almost 'classical' antirheumatic treatment. Cytostatic and cytotoxic drugs differ in how they influence nuclear metabolism thus preventing cell growth. We consider the following drugs as immunosuppressive, although others used in antitumor therapy could be added but are too toxic to use in rheumatic diseases: cyclophosphamide and chlorambu-

[1] Dedicated to Prof. Dr. *H.J. Dengler* on the occasion of his 60th birthday.

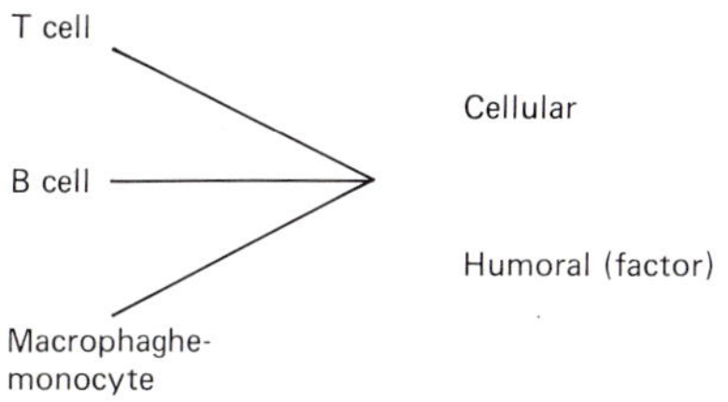

Fig. 1. Cellular and humoral partners of immune response.

Table I. Immunosuppressants

Humoral	Cellular
Plasmapheresis	Thoracic duct drainage Lymphapheresis Antilymphocyte serum T suppressor cell activation Cytostatic drugs Radiation

cil as representatives of the alkylating agents, azathioprine and 6-mercaptopurine as purine analogs, and the folic acid antagonist methotrexate (table II). More recent developments such as levamisole and cyclosporin A act as immunosuppressants by more selectively influencing T cell subsets: levamisole seems to augment T suppressor cell function whereas cyclosporin A works via reduction of T cytotoxic cells. The former is able to influence the progression of RA but exhibits some toxic side effects such as agranulocytotosis, the latter is widely used in the suppression of transplantation immunity, its activity in RA treatment is disappointing. The newly developed cyclosporin D, on the contrary, has some influence on RA but none on transplantation immunity (table III).

The use of cytostatic drugs increases the risk of side effects (table IV), whereas the risk of developing lymphoma is well documented for several immunosuppressants such as azathioprine, cyclophosphamide and chlorambucil. Some of these side effects need the discontinuation of the drugs, others only reduction of the daily dose. Bone marrow suppression needs continuous monitoring of peripheral blood cells. A

Table II. Cytostatic/cytotoxic drugs

Alkylating agents	Purine analogs	Folic acid antagonists
Cyclophosphamide	azathioprine	methotrexate
Chlorambucil	6-mercaptopurine	

Table III. Activity on T cell subsets

	Transplantation	RA
Levamisole	–	+
Cyclosporin A	+	–
Cyclosporin D	–	+ (?)

Table IV. Side effects of cytostatic agents

	Folic acid antagonists	Purine analogs	Alkylating agents
Bone marrow suppression	+	+	+
Susceptibility to infection	+	+	+
Hepatotoxicity	+	+	0
Infertility	0	0	+
Carcinogenicity	0	+	+
Cystitis	0	0	+
Mucous membrane lesions	+	0	0

fall in total lymphocyte count below 500/mm^3 correlates with an increased susceptibility to bacterial and viral infections. Hepatotoxity is an individual effect and rather nonpredictable. Specific side effects occur under the use of cyclophosphamide, such as cystitis which even might develop to carcinoma of the urinary bladder – mucous membrane lesions including ulcers and bleeding may result from use of methotrexate (table IV). Most of these side effects disappear on withdrawal of the drug.

More serious is the problem of onconeogenesis. There are well-documented risks of tumor aquisition, solely of the lymphoma type, by the use of azathioprine, cyclophosphamide and chlorambucil. These side effects seem dose dependent, time dependent and drug dependent. Use of chlorambucil in RA patients at a daily dose of 0.2 mg/kg increases the incidence of acute leukemia from 0.1 to 1% in a French population. If, on the other hand, the total dose does not exceed 1 g and the duration of treatment 6 months, there is no increased risk of leukemia. It nevertheless reduces the treatment with these drugs to those cases which exhibit all the features of highly active and progressive disease.

There are some patients treated with immunosuppressants, where the induction of remission occurs at a dose where none of the parameters of immunosuppression, such as reduced lymphocyte counts, usually occur. It is therefore possible that the activity of these drugs include mechanisms which cannot be detected by the general measures of immunosuppression. It is thinkable that only minimal changes in a specific immune response are able to restore a normal response thus extinguishing the disease. In this respect, new therapeutic trials should include those which either use low-dose immunosuppressants or more pulse-like therapies.

Prof. Dr. I. Stroehmann, Röntgenstrasse 6, D-5300 Bonn 2 (BRD)

Rheumatology, vol. 9, pp. 46–50 (Karger, Basel 1986)

20 Years of Methotrexate in the Treatment of Autoimmune Diseases

P.A. Miescher

Département de Médecine, Hôpital Cantonal Universitaire de Genève, Genève, Switzerland

The antimetabolites 6-mercaptopurine (6-MP) and methotrexate (MTX) have both been reported to inhibit delayed hypersensitivity in the guinea pig [1–3], but only MTX suppressed antibody formation in this animal [2]. With the possibility in mind of using MTX for patients with autoimmune diseases, we investigated, with *Spiegelberg*, the effectiveness of the two antimetabolites in the treatment of experimental autoimmune thyroiditis in guinea pigs [7]. The results of these investigations showed that these antimetabolites do not only prevent the development of thyroiditis, but also have an effect once the disease is established. MTX appeared more effective in terms of toxicity/efficacy ratio.

Impressed by these results, we introduced, in 1964, methotrexate in the treatment of a patient suffering from acute systemic lupus erythematosus (SLE) with high temperature and convulsions (fig. 1). A single injection of 100 mg MTX i.v. was followed by a rapid normalization of the body temperature, cessation of convulsions, and a rise in the number of platelets from 48,000 to 120,000. A second injection of 50 mg 1 week later was also well tolerated although the platelet number fell subsequently to 54,000/mm^3. Following this observation, we treated patients suffering from chronic immunological conditions with weekly injections of MTX at a dosage of 25 mg per injection. Encouraging results were obtained with patients suffering from acute SLE as well as with patients suffering from chronic cutaneous vasculitis (fig. 2). However, a few patients developed signs of toxicity with nausea and mouth ulcers after 6–8 weeks of treatment.

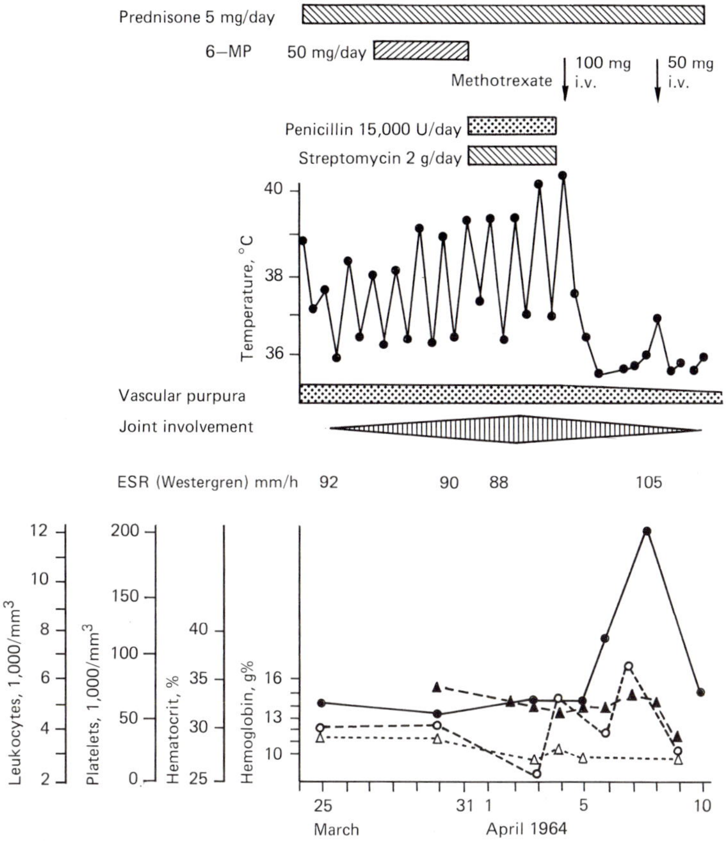

Fig. 1. Effect of i.v. MTX on a highly febrile course of SLE in a 34-year-old female patient.

We then tried MTX given orally at a dose of 3 × 2.5–3 × 5 mg of MTX per week. Oral administration proved less successful in patients with acute diseases but was found suitable for maintenance therapy.

Since patients suffering from chronic autoimmune diseases usually require prolonged treatment, we have adopted a strategy of combining various drugs in an attempt to obtain disease control with a minimum of side reactions. In addition to steroids, the two antimetabolites, 6-MP and MTX, were found to be suitable for this purpose. Furthermore, it has been our experience that prolonged (1–2 years)

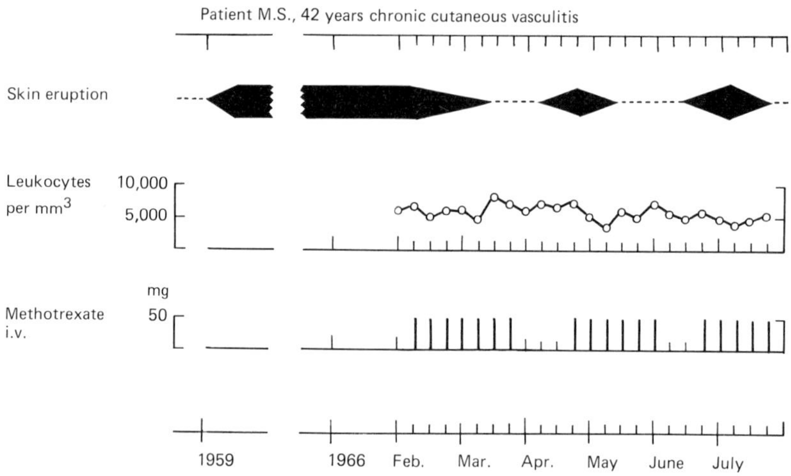

Fig. 2. Remission induction therapy of patient M.S. suffering from chronic cutaneous vasculitis.

administration of 6-MP (or azathioprine) produces a state of resistance which is not only reflected by lack of responsiveness to but also by increased tolerance of large doses of this antimetabolite. On the other hand, MTX has been shown to lead to toxicity when given on a continuous basis. For this reason, we have used for maintenance therapy the two drugs on an alternate month schedule, as illustrated in the case reported in figure 3. While in most conditions these antimetabolites do not suffice to control disease activity when given without steroids, some patients with chronic cutaneous vasculitis did not require steroids. The drug-dose dependence of the responsiveness to treatment is seen in patient M.S., first with the remission induction therapy of i.v. MTX (fig. 2), subsequently during maintenance therapy. When, on two occasions, the azathioprine dose was lowered from 100 to 50 mg/day, the skin rash reappeared each time. During the month of MTX medication following the month of low-dose azathioprine, the skin rash disappeared again thus showing the efficacy of MTX. The alternate month schedule sometimes also permits a comparison of the two antimetabolites. In 1 patient suffering from a severe form of dermatomyositis, the clinical condition improved markedly during each MTX

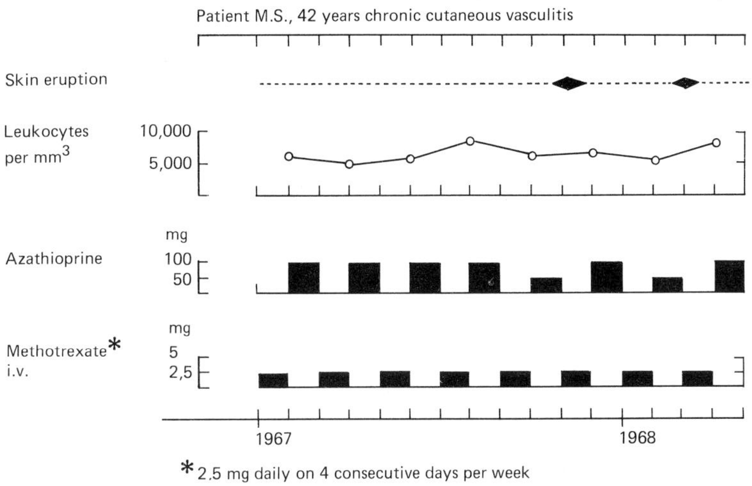

Fig. 3. Maintenance therapy in patient M.S. (see fig. 2) using MTX and azathioprine on an alternate-month schedule.

month but not during the 6-MP medication months. However, converse behavior has also been observed [4–6].

More recently, we have used i.v. MTX in combination with steroids and cyclosporin (CyA) for remission induction in patients with severe rheumatoid arthritis resistant to conventional forms of treatment. The following treatment schedule has led to a surprisingly rapid improvement of the inflammatory joint and general inflammatory condition in 6 patients: (1) 15 mg of MTX i.v. once per week together with 3/4 liters of tea to ensure on-going diuresis (administered on days 1, 15, 18 and 22); (2) fluocortolone, 15 mg, 4 times per week; (3) CyA, 5 mg/kg, 6 days per week.

This drug combination is interesting since none of the various compounds is of significant early efficacy if used alone. Indeed, the dose of steroids is too small to be effective in these severe cases. The effect of MTX is not very evident after 4 weeks of treatment and CyA in association with steroids only proved beneficial after about 6 weeks' treatment. However, all 6 patients already showed marked improvement after 3–4 weeks of treatment. After the first month, patients were maintained on a combined fluocortolone/CyA treatment for an additional 3 months. We then alternated CyA on a monthly basis with azathioprine (100 mg/day, 6 days per week). Whenever patients experienced an exacerbation of their RA, an additional series of 2–4 i.v. injections of MTX together with CyA was shown to be effective. For instance, in 1 patient who contracted a severe form of influenza with subsequent exacerbation of his joint condition, swelling of joints subsided after the second injection of MTX.

Many clinicians have been afraid to use MTX in view of its relatively small therapeutic margin. It is formally contraindicated in patients with an impaired kidney function. After the impressive result obtained in our first SLE patient (fig. 1), we gave a trial dose of 2.5 mg MTX i.v. to another patient with a creatinine clearance of 20 ml/min. This small dose produced nausea sufficiently severe to preclude further use of the drug in patients with impaired kidney function. Patients with liver damage also do not tolerate MTX. However, bearing these restrictions in mind, we have widely used MTX in our patients for the past 20 years.

References

1 Hoyer, J.; Condie, R.M.: Suppression of delayed hypersensitivity in guinea pigs and rabbits with 6-mercaptopurine. Fed. Proc. *21:*277 (1962).

2 Friedman, R.M.; Buckler, C.E.; Baron, S.: The effect of aminomethylpterylglutamic acid on the development of skin hypersensitivity and on antibody formation in guinea pigs. J. exp. Med. *114:*173 (1961).

3 Genghof, D.S.; Battisto, J.R.: Antibody production in guinea pigs receiving 6-mercaptopurine. Proc. Soc. exp. Biol. Med. *107:*933 (1961).

4 Miescher, P.; Riethmüller, G.: Diagnosis and treatment of systemic lupus erythematosus. Semin. Hematol. *2:*1 (1965).

5 Miescher, P.A.; Gerebtzoff, A.; Lambert, P.H.: Immunosuppressive therapy; in Miescher, Proc. Int. Wiesbaden Symp., 1972, pp. 13–48 (Schwabe, Basel 1973).

6 Miescher, P.A.; Beris, P.: Immunosuppressive therapy in the treatment of autoimmune diseases. Springer Semin. Immunopathol. *7:*69 (1984).

7 Spiegelberg, H.L.; Miescher, P.A.: The effect of 6-mercaptopurine and methotrexate on experimental immune thyroiditis in guinea pigs. J. exp. Med. *118:* 869 (1963).

Prof. Dr. P.A. Miescher, Département de Médecine, Division d'Hématologie et Centre de Transfusion Sanguine, Hôpital Cantonal Universitaire de Genève, 25, rue Micheli-du-Crest, CH–1211 Genève 4 (Switzerland)

Rheumatology, vol. 9, pp. 51–53 (Karger, Basel 1986)

Discussion 2

Michaels: I have just a comment about what Dr. *Stroehmann* has said about methotrexate and carcinogenicity. I believe the longest experience with the low-dose methotrexate is in patients with psoriasis or psoriatic arthritis. In 1975 the experience of the Cleveland Clinic with the incidence of tumors in patients with the long-term treatment of methotrexate for psoriasis was published. Basically, the paper said that there was no increased incidence of carcinomas or lymphomas above what would be expected in the general population.

Stroehmann: I just want to give you a short reply. You remember that chlorambucil is used a long time already in rheumatoid arthritis. But convincing reports on its toxicity in terms of carcinogenicity have appeared, let's say, only in the mid-1970s. You must admit that an increase of the incidence of acute leukemia from 0.1 to 1% might not be detected in a small study. If I remember, the Paris group has reported 3,800 cases to detect the increased rate of acute leukemia.

Wilke: There was another paper about the treatment of trophoblastic tumors with methotrexate. This isn't as long term as what we are talking about for rheumatoid arthritis, but they had 8 years follow-up and something like 400 patients and found no increased incidence. In fact, it looked as if those patients had a lower incidence of lymphoma than did the population as a whole. This is good information, too.

Weinstein: It didn't take that long to discover that cytoxine did cause an increased incidence of lymphoma in patients with rheumatoid arthritis. In fact, I think, the Tucson group which espoused the use of cyclophosphamide then abandoned it within a few years because of the marked increase in incidences of lymphoma in rheumatoid arthritis patients. So methotrexate does appear to be different. If it's going to happen it's going to take much longer.

Rau: In a Scandinavian study with some 30,000 patients, the incidence of malignancies in patients with rheumatoid arthritis treated with cytotoxic agents was not higher than in a population of RA patients without cytotoxic treatment. On the other hand, the incidence of lymphoma in RA patients was higher than in the general population.

Stroehmann: Yes, the incidence of lymphoma in RA patients is about 10% more than in the normal population. But the risk for example with chlorambucil is about 10 times higher.

Furst: I remember an article in 1978 on polycythemia vera and chlorambucil. In this study there were only 400 or 500 patients, and within 3 years they stopped the study simply because they found AML after chlorambucil. Again I think that if methotrexate has that effect, it's going to be less and is going to be very much delayed.

Michaels: One other representative might be Dr. *Roenigk* who reviewed the guidelines for methotrexate therapy for dermatologists in 1982.

He looked back at the 20 plus years of experience with the use of methotrexate in psoriasis and found no increased incidence in patients treated with long-term methotrexate.

Jürgens: If you compare the mechanism of action, there is no evidence with this type of drug that you can compare it with the other drugs we are told about. Because you have an inhibition of an enzyme which has a special lifetime and then the enzyme gets back. And I can't imagine how there may be a mechanism of starting any changes in the DNA or RNA.

Stroehmann: If changes in DNA or RNA were the essential effect, you would expect a certain rate of carcinogenesis respecting all tumors. But it is a fact, that it is only lymphoma, and we believe now that this is not a question of that you transfer something in the DNA or change the DNA but this is probably really something which has to do with the immune surveillance. So, most probably, it is a question of immunosuppression, and not of a typical action on the DNA background.

Furst: I was fascinated by Dr. *Miescher's* work. A good number of years ago now, we used a combination of immunosuppressives. It is the wheel's turn, we are beginning to use it again. And I wondered, referring back to Dr. *Jürgens'* slides before, that he said that there was some sort of synergy between 6-MP and methotrexate, and how that might apply, whatever the mechanism might be, with respect to Dr. *Miescher's* work.

Jürgens: This co-synergy between methotrexate and 6-mercaptopurine is based on the synergistic attack on the purine metabolism of both compounds. There are reports from the National Cancer Institute that both drugs actually are synergistic in reducing the storage pool for particular compounds in purine metabolism, but in order to establish the full synergistic effect, you have to have high 6-mercaptopurine levels followed by methotrexate and you are not likely to obtain those levels by giving 6-mercaptopurine p.o. So, I'm not sure that there is evidence enough to suggest that p.o. administration of both drugs will result in any synergistic effect.

Weinstein: I was also interested in Dr. *Miescher's* work, but I am also a little disturbed. The use of cytotoxic agents and certainly combinations of cytotoxic agents in diseases like lupus without controlled trials is disturbing. I wonder, if you have done any randomized studies to show that what you are doing is in fact worthwhile and better than steroids alone.

Miescher: This is a very pertinent question. How can we prove what we are doing in our patients with lupus. I think Dr. *Decker* once mentioned the following thing: if we are analyzing azathioprine, it cannot replace steroids. If we analyze drug after drug, they are not doing very much when used as single drugs in controlled studies. Yet, if we are comparing mortality in lupus, 30 years ago and today, it's totally different. We have to do a properly controlled study, but it is simply impossible. Because, as I have mentioned, every lupus patient is slightly different from another one. My colleagues tell me, why should you use methotrexate or azathioprine. We can do the same thing with steroids. You are right, you can do the same thing with steroids, if you do not ask about the overall well-being of your patients. And if I speak about side reactions, when I have visitors, they ask me: but we see almost no Cushings syndrome in your waiting room. What is the reason. The only way we can treat our patients with a minimum amount of subjective side reactions, moon face, and so on, is by tapering down steroids and replacing steroids with immunosuppressive agents, which have different side reactions. Not all patients are treated from the beginning. Some patients come with large amounts of steroids necessary to maintain, I would say, or to prevent disease activity. And for this reason, to answer your question, we do not have a controlled study. But we can treat patients without exposing those patients to the well-known side reactions you see in other lupus clinics.

Weinstein: You don't see any avascular necrosis in your lupus patients?

Miescher: Now, this is another important question. I agree, we are not discussing steroids today, but, for instance, for the past 10 years, we have never given steroids on a daily basis. If you put patients on daily steroid medication, you get aseptic necrosis of the bones, you get the typical skin lesions, you get very, very severe side reactions, which we do not see if you introduce at least 2 days per week without steroids, preferentially even 3 days, but 2 days is the minimum amount. Now this we have shown by checking the adrenal function with 0.25 mg of ACTH. And those patients who are on long-term steroid medication with two nonconsecutive days per week without steroids, have a practically normal stimular ability of the adrenals which means they can face stress. So in other words again, it's not only the question of methotrexate or 6-MP, it's also the question of how to give steroids to your patients.

Stroehmann: I'm aware of quite a few studies concerning SLE and treatment with cytotoxic drugs and/or cortisone. All these studies have been randomized and controlled. And they have shown at least one thing. There is a major advantage for cyclophosphamide. This is for sure, and then comes chlorambucil and it has also been shown, at least in these studies, that azathioprine is not effective, not more effective than cortisone alone in the treatment of SLE. And there is another study which shows that azathioprine alone or in combination with cortisone is not effective in the treatment of Wegener's granulomatosis, where it has been shown that the absolute drug which brings the remission is the combination of cortisone plus cyclophosphamide.

Miescher: I would only say to this point: since we are not dealing with a homogenous control population, you may give the wrong treatment based on a pseudo-statistical analysis. All I can tell you, I avoid in lupus patients as much as possible alkylating agents, just because of the high level of cancerogenicity. If I'm using alkylating agents, it is only in terms of pulse therapy, which is very useful.

Warnatz: Dr. *Miescher,* are you really convinced that you are dealing with immunosuppression?

Miescher: It is not easy to demonstrate in realistic terms that you are applying immunosuppressive treatment. You can do it with high dosages of steroids. If you give to patients with severe kidney involvement for 6 months a high dose of steroids, in our schedule always leaving 2 days without applying steroids, you get a severe hypogammaglobulinemia. Otherwise it is very difficult, because lupus activity goes together with lymphopenia. You can not take the lymphocyte count as evidence of immunosuppression. I would say, one of the most trivial examples that we are doing immunosuppression with a relatively low dose of steroids together with azathioprine and methotrexate is that the incidence of herpes zoster is much higher in those patients, it is very high. Also common warts are much more prominent in those patients. We have the clinical expression of a mild immunosuppression. But to objectively demonstrate that the effect we are achieving is a result of immunosuppression is impossible.

Williams: Thank you for your discussion.

Rheumatology, vol. 9, pp. 54–58 (Karger, Basel 1986)

Mode of Action of Methotrexate in Rheumatic Diseases

Robert F. Willkens

Harborview Medical Center, University of Washington, Seattle, Wash., USA

The primary action of methotrexate (MTX) is as an inhibitor of dihydrofolate reductase. The consequence of this inhibition is interference with S-phase maturation of cells with all the consequences contingent upon that cell's maturation inhibition. It is unlikely that the action of MTX in rheumatic diseases is different from this primary action on cells. The absence of a more precise understanding relates not only to the failure of awareness of the ultimate pathophysiology of the rheumatic diseases, but also to the distinction between normal and abnormal cell maturation and disease.

Page et al. [1] demonstrated an anti-inflammatory response of 6-mercaptopurine and other antimetabolites against mononuclear cell response in rabbits. This impairment was not related to granulocyte exudation. Curiously, they were unable to show such an impaired response for aminopterin, another folate antagonist.

Subsequently, *Hersh* [2] showed an inhibitory effect of monocyte migration by MTX using a Rebuck-window technique. In this model, patients about to be treated for leukemia with chemotherapeutic agents underwent forearm abrasion followed by application of a coverslip. No further local irritant was applied. An initial immediate polymorphonuclear response was followed by a monocyte migration after 4 h which continued to 48 h. Inhibition of this response was studied and was found to be different depending upon the therapeutic regimen applied. The median 'on therapy' value at 4, 8, and 24 h were 0, 8.5, and 23%, as opposed to control values of 4, 24.5 and 86% ($p < 0.001$). Inhibition of monocyte cell exudation did not correlate with white blood cell counts and the granulocyte count in the peripheral blood. This study established that MTX therapy inhibited mononuclear cell migration at experimental inflammatory sites in man. The fact that rabbits had failed to show such an effect probably reflects species differences in that MTX does not affect the development of antibodies to BCG on sheep erythrocytes in rabbits but does in other species.

Stevens and Willoughby [3] tested a spectrum of chemotherapeutic agents in models of acute and chronic inflammation. Thermal injury and turpentine peritoneal injury were the models used. MTX was administered intraperitoneally over a 3-day period or by a single injection before injury. There was some difference in response of cells to these different acute models, but this inhibition of cells was once again more apparent on sensitive mononuclears; the authors concluded that the anti-inflammatory effect correlated with the degree to which MTX lowered the number of lymphocytes in the peripheral blood.

More recently, *Zeller and Gudewiz* [4] injected rats intraperitoneally with either MTX or ethanol and harvested peritoneal exudate cells after 96 h. These cells were characterized and glucose oxidation assessed by a $^{14}CO_2$ collection method. Cell differential counts indicated that peritoneal macrophage accumulation is selectively depressed, while polymorphonuclear leukocyte population is enriched. The cell viability of peritoneal macrophage cells was unaffected although glucose oxidation was depressed as a function of MTX dose.

This study by *Zeller and Gudewiz* not only confirms the selective effect of MTX on mononuclear cells, but also when comparing this effect to the effect of MTX on lymphoma cells, normal cells were found to be more resistant than tumor cells.

Mansour et al. [5] evaluated the proliferative responses of rat lymphocytes and spleen cells to phytohemagglutinin (PHA). Following intraperitoneal injection, lymphopenia occurred with all doses when counts were performed at 24 h after the last of four consecutive daily doses. At most doses used, MTX impaired peripheral blood lymphocyte PHA responses but enhanced spleen cell responses. These authors were concerned with the discordant effects of MTX on different mononuclear cells. This study confirms a different effect on peripheral blood lymphocytes and spleen cells by MTX and establishes an enhanced incorporation of tritiated thymidine by exposed lymphocytes at some doses of MTX.

The store of reduced folate in spleen versus peripheral blood lymphocytes may partially explain the lesser sensitivity of spleen cells to MTX. The enhanced incorporation of thymidine may relate to alterations of the internal metabolism of the cell. More generally, one would consider that it relates to the network of cellular interaction; while inhibition of certain cells may have an enhancing effect, others may have a compensatory depressing effect.

As more is learned about the clinical effects of MTX, we recognize the importance of cell permeability at different free and protein-bound drug levels. The accumulation of polyglutamates is another of these factors that influences the effect of MTX in vivo. Intercellular folates exist as variable-length polyglutamates. These gammapeptide linked forms of folate have unique properties, i.e. polyglutamates dissociate from dihydrofolate more slowly than MTX and may thus be less reversible in their inhibition of the enzyme.

These studies suggest that antimetabolites and chemotherapeutic agents have effects that could be called anti-inflammatory. This designation may be more somatic than scientific. After all, whether a drug is called anti-inflammatory, immunoregulatory, or cytotoxic more often reflects the experimental system used to demonstrate the effect and/or the dose of the drug used, than insight into its actual effect.

Goodwin and Ceuppens [6] have performed a number of experiments both in vitro with lymphocytes, and in vivo with rheumatoid patients. In these experiments they have demonstrated an inhibition of rheumatoid factor production. Isolated lymphocytes were less influenced by nonsteroidal anti-inflammatory drugs, such as Indocin, than was the cascade of monocyte suppressor-helper B cells. Their hypothesis is that prostaglandins serve as a stimulus to suppressor cells which are turned off by PGE_2. Since PGE_2 is felt to have its origin from lymphocytes stimulated by a number of antigens, including immune complexes, the prostaglandin effect may be considered to be a negative feedback response.

The effect of MTX on the primary immune response [7] was assessed in leukemic patients by characterizing the development of antibodies to *Escherichia coli* antigen and tetanus toxoid during therapy. Complete suppression of the antibody response was noted during therapy and half the patients never developed antibody even after therapy was discontinued. The remaining patients did develop mercaptoethanol-sensitive antibody after a latent period that averaged 12 days.

Kazmer et al. [8], evaluating the effect of immunoregulatory drugs on purine biosynthesis of a human lymphoblast cell culture, noted a profound cytotoxic effect of MTX on B cells. This effect was exceeded in these experiments only by arabinofuranosylcytosine (Ara-C). Other studies have shown an inhibitory effect of MTX on delayed hypersensitivity. The inhibitory effect of MTX on dihydrofolate reductase preventing the formation of DNA precursors is the primary mechanism of its actions; its reversibility by the restoration of the pool of reduced folate and by leukovorin factor makes it an attractive moderator of cell maturation and development.

At the 1984 American Rheumatism Association meeting in Minneapolis, *Anderson* et al. [9] reported their evaluations of peripheral blood lymphocytes from rheumatoid patients treated for RA with MTX. They found no changes in mononuclear cell responses to PHA or PWM. There were changes in T/B cell numbers but there were no changes in the numbers of T cell subsets. Rheumatoid factor and immunoglobulin levels were also unchanged.

Despite an inability to characterize accurately the site or mechanism of action of MTX in RA, it seems most likely that MTX is an antimetabolite and an immunosuppressive and cytotoxic agent, depending upon dose used. Furthermore, MTX probably has a prominent anti-inflammatory action related to dose and timing of administration.

References

1 Page, A.R.; Condie, R.M.; Good, R.A.: Clinical studies on the anti-inflammatory activity of 6-mercaptopurine. Blood *20:* 118 (1962).

2 Hersh, E.M.; Wong, V.G.; Freireich, E.J.: Inhibition of the local inflammatory response in man by antimetabolites. Blood *27:* 38 (1966).

3 Stevens, J.E.; Willoughby, D.A.: The anti-inflammatory effect of some immunosuppressive agents. J. Path. *97:* 367 (1969).

4 Zeller, J.N.; Gudewiz, P.W.: Modulation of inflammatory peritoneal cell function and metabolism of methotrexate. J. Immunopath. *3:* 309 (1981–82).

5 Mansour, A.; Henderson, J.A.; Nelson, D.S.: Effect of methotrexate on the response of rat lymphocytes to phytohaemagglutinin. Clin. exp. Immunol. *34:* 393 (1978).

6 Goodwin, J.S.; Ceuppens, J.L.: Effect of nonsteroidal anti-inflammatory drugs on immune function. Semin. Arth. Rheum. *13:* 134 (1983).

7 Mitchell, M.S.; Wade, M.E., DeConti, R.C.; Bertino, J.R.; Calabresi, P.: Immunosuppressive effects of cytosine arabinoside and methotrexate in man. Ann. intern. Med. *70:* 535 (1969).
8 Kazmers, I.S.; Daddona, P.E.; Dalke, P.; Kelley, W.N.: Effect of immunosuppressive agents on human T and B lymphoblasts. Exp. Pharm. *32:* 805 (1983).
9 Anderson, P.A.; West, S.C.; Claypool, R.C.: Pulse methotrexate therapy for rheumatoid arthritis: A double-blind crossover study. Arthritis Rheum. *27:* suppl., p. S25 (1984).

Robert F. Willkens, MD, Harborview Medical Center, University of Washington, 325 Ninth Avenue, Seattle, WA 98104 (USA)

Rheumatology, vol. 9, p. 59 (Karger, Basel 1986)

Discussion 3

Fenner: Is there any evidence that the concentration of the drug in the synovial fluid is important for this special activity in rheumatoid arthritis?

Willkens: I don't know if anyone measured the synovial fluid concentrations of the drug.

Michaels: I've been impressed and, I think, probably many other people in this room have also been impressed with the speed of clinical response that we see with methotrexate compared to the other agents that we use. We know that methotrexate is important in methylation reactions as discussed earlier today. Do we know of any phlogistic agent that is highly dependent on methylation reactions for its activity or any cell that is highly dependent on methylation reactions or its viability or existence that methotrexate might be affecting so rapidly?

Willkens: The only thing I could think of would be because of its rapid generation of the groups and further than that I don't know.

Furst: We have had some very preliminary data that indicated the possibility, and note how carefully I'm saying all this, that there is a decrease in T helper function in 3 patients.

Weinstein: Well, we have very preliminary evidence in normal rats given low-dose methotrexate that T helper cells are decreased at low doses, in the doses similar to be given to man.

Stroehmann: Well, you should also consider that the tests we do in the lab to determine T helper or T suppressor cells are really dependent on the type of method we use.

Furst: I would agree with that. That is why I was being extremely careful with what I had to say.

Rau: Well, I think that the effect of methotrexate on the inflammation is so rapid, if you give the appropriate dose, that you can't explain this effect with an immunosuppressive way of action and therefore I think there must be a nonspecific, anti-inflammatory effect, perhaps a cytotoxic effect, on all cells which contribute to the inflammation.

Wilke: One comment, quite a few steroids in low dose are mostly anti-inflammatory. Quite a few steroids in high dose are immunosuppressive. We may not be talking about one effect but many, many effects with methotrexate.

Rheumatology, vol. 9, pp. 60–87 (Karger, Basel 1986)

Methotrexate Therapy in Psoriasis and Psoriatic Arthritis

A Review

A. Nyfors

Department of Dermatology, Haukeland University Hospital, Bergen, Norway

Introduction

The author received his dermatological training in the Department of Dermatology, The Finsen Institute, Copenhagen, from 1967 to 1978 and confronted psoriasis vulgaris about 40,000 times partly during daily routine work and partly by building up and following a group of 248 psoriatics in long-term methotrexate (MTX) therapy [66]. This paper aims to be an overview on the subject rather than an extensive review. It has been written to be useful to rheumatologists and dermatologists. When nothing else is mentioned in this article psoriasis means ordinary psoriasis of the skin, i.e. psoriasis vulgaris. The subject will be dealt with in a series of sections, the title of which will be seen in the following as section heads.

Definition of Psoriasis vulgaris

An exact definition of psoriasis can not be given, inter alia because its etiology is unknown. Psoriasis may be described as a chronic skin disease manifesting itself in pink, scaly, confluent papules [7]. It is a common skin disease as about 2% of the population in Western Europe develop it. Recent reviews on psoriasis have been written by *Nyfors* [67], *Cram* [19], *Champion* [14], *Anderson* [1], and *Farber and Cox* [24].

Definition of Psoriatic Arthritis

Psoriatic arthritis is a polyarthritis in patients having psoriasis vulgaris. They are theoretically negative for rheumatic factor and Rose Waaler test. The common sites of the arthritis are the distal interphalangeal joints of the fingers and toes, next in order the sacroiliac joints, more rarely the wrist, ankles and hip joints. Psoriasis arthritis occurs in 2–7% of patients with psoriasis vulgaris. *Moll* has reviewed the clinical spectrum of psoriatic arthritis [56] as well as the seronegative arthropathies [57].

Course and Prognosis of Psoriasis vulgaris

The natural history of psoriasis is lifelong in the majority of patients. In 50% of cases psoriasis sets in before the age of 13 [53, 64]. In the age range 50–70 the disease seems to wane [106]. By the available methods of treatment in the 1950s and early 1960s, viz. tar, dithranol, light baths, many psoriatics had to spend months admitted in dermatological departments. The need for faster, cleaner and easier treatment was great in the 1950s.

History of MTX Therapy in Psoriasis

Folic acid (pteroylglutamic acid) was identified and synthesized in 1946 by a research team from the American Cyanamid Company [2]. In the same year *Farber* et al. [25] tried treating 11 patients with acute leukemia with folic acid analogues (pteroyldiglutamic and pteroyltriglutamic acid 10–30 mg daily) contained in splenic extracts and raw liver on the basis of findings by *Leuchtenberger* et al. [51] that such extracts (*Lactobacillus casei* factor) inhibited the development of tumors in mice. In *Farber's trial,* however, these folic acid analogues aggravated the course of acute leukemia [25]. Thereafter, the production of compounds possessing antifolic acid properties was instituted.

In 1947, it was demonstrated that crude x-methyl folic acid (1–10 g/kg diet) inhibited the growth of rats and depressed their blood counts and bone marrow, most profoundly the total white cell and granulocyte counts [30]. *Seeger* et al. [94] synthesized the folic acid antagonist aminopterin, chemically 4-amino folic acid and created by substituting the OH group by an amino group in position 4 in the folic acid molecule.

Farber et al. [26] succeeded in obtaining temporary remission of leukemia in 5 children by daily injections of 0.5–1.0 mg aminopterin. In the search for an agent which was less toxic than aminopterin, *Smith* et al. [97], produced in 1948, 4-amino-10-methyl-folic acid, amethopterin, methotrexate (MTX).

In 1949, it was demonstrated by *Franklin* et al. [31] that in animal experiments MTX in a diet with 0.3–1.0 mg MTX/kg diet was less toxic than aminopterin (did not retard growth and caused little alopecia).

Aminopterin was introduced in the treatment of psoriasis in 1951, when *Gubner* et al. [36] found clearing of psoriasis in a patient being treated with 2 mg aminopterin for rheumatoid arthritis.

Rees et al. adopted this treatment for psoriasis, in daily doses of 0.5–1.0 mg aminopterin, and in publications from 1955 [76] and 1959 [77] they reported that it had resulted in clearing or great improvement in 80% of 171 and 229 patients, respectively.

Edmundson and Guy [23], studying 62 psoriatics, found MTX to be equally effective and less toxic than aminopterin, in oral doses of 2.5 mg MTX and 0.5 mg aminopterin, respectively. This experience with regard to toxicity was confirmed in 1961 by *Rees and Bennett* [78].

To decrease side effects, *Roenigk* et al. [81], at the suggestions of *Callaway*, and *Callaway* et al. [12] recommended treating psoriasis with one single weekly oral dose of 25 mg MTX. In 1971, *Weinstein and Frost* [107] suggested dividing the weekly oral dose of MTX into three portions, given at intervals of 12 h to prolong the effect of MTX treatment upon the proliferating epidermal cells in psoriasis.

The history of MTX has been reported especially by *Jukes* [44] and as regards psoriasis by *Nyfors and Brodthagen* [63] and *Rees* [80].

Methotrexate

It falls outside the frame of this paper to go into details with the pharmacology of this folic acid antagonist, MTX. Interested readers are referred to reviews such as *Bleyer* [9], *Shen and Azarnoff* [95], *Nyfors* [66], *Schornagel* [91], *Jolivet* [43], as well as *Jürgens* [45].

Indications

The indications for MTX therapy of psoriasis are described in the MTX guidelines by *Roenigk* et al. [84]: The decision to administer MTX for the therapy of psoriasis should be individualized. Each patient should be evaluated with reference to disease severity, amount of discomfort, degree of incapacity, and general medical and psychological condition. MTX is indicated in the symptomatic control of recalcitrant psoriasis not responsive to topical therapy. The diagnosis of psoriasis and the need for MTX therapy should be established by dermatologic consultation.

In general, these patients should have severe psoriasis which may be life-ruining physically, emotionally, or economically. The psoriasis should be so severe that it cannot be adequately controlled by standard topical antipsoriatic therapy. Some examples of candidates for MTX therapy are. (A) psoriatic erythroderma; (B) psoriatic arthritis; (C) acute pustular psoriasis – Von Zumbusch type; (D) localized pustular psoriasis; (E) psoriasis in areas of body preventing economic employment, and (F) extensive psoriasis.

MTX has also been used widely in rheumatic arthritis [39, 108, 109].

Contraindications

The MTX guidelines [84] list the following relative contraindications:

A	Significant renal function abnormalities.
B	Significant liver function abnormalities.
C	Pregnancy – near absolute!
D	Men or women in fertile age – appropriate steps should be taken to avoid conception during and for at least 12 weeks after discontinuing MTX therapy.
E	Hepatitis – active or recent.
F	Cirrhosis.
G	Severe anemia, leukopenia, or thrombocytopenia.
H	Peptic ulcer, active – near absolute!
I	Excessive alcohol consumption.
J	Active infectious disease (i.e. tuberculosis, pyelonephritis).
K	Unreliable patient.

Circumstances may arise in which contraindications must be waived, i.e. when benefits can be expected to outweigh the risks of MTX therapy in the individual patient.

Pre-MTX Evaluations

The following pre-MTX evaluations of the psoriatic is recommended in the MTX guidelines [84]:

	History and physical examination
	Laboratory
1	Hematologic
	Hemoglobin
	Leukocyte count
	Platelet count
2	Renal function studies
	Urinalysis
	Blood urea nitrogen (BUN)
	Serum creatinine or creatinine clearance
3	Liver function studies
	Serum glutamic puruvic transaminase (SGPT)
	Serum glutamic oxaloacetic transaminase (SGOT)
	Alkaline phosphatase and serum bilirubin
4	X-ray film of chest

Liver Biopsy

A thorough history and physical examination will identify some patients with preexisting risk factors, such as liver disease. Chemical liver function tests will identify only some cases of existing liver disease. Quadratic multiple ordered laboratory tests have been able to differentiate between alcoholic and nonalcoholic patients with different degrees of liver disease [89].

Start of MTX Treatment for Psoriasis

Before starting taking the pre-MTX laboratory tests and performing the liver biopsy the patient must make up his mind, whether he is able/willing to live with an alcohol intake of 0–3, say maximum 5 drinks a week. If the patient is not willing to abstain from alcohol or cannot do so because of much representation as part of his job, I would not start the patient on MTX therapy [69]. Quite a few patients lie about their alcohol intake. The worst liar in my group of patients [66] was a catholic minister, whom loneliness drove into alcoholism. He was such a good liar as to hide his abuse from us until a liver biopsy disclosed alcoholic hepatitis and cirrhosis. The implications of alcohol intake will be discussed in the review on hepatotoxicity of MTX therapy in psoriasis [69].

When the results of the pre-MTX evaluation allow us to begin MTX therapy a pre-MTX liver biopsy should be performed unless it is contraindicated (a low platelet count, abnormally high serum bilirubin or a recent coronary infarct, to give some examples). The purpose of the pre-MTX liver biopsy is to have a liver histology to compare later liver histology findings during MTX therapy with [69].

There are different opinions whether to begin MTX therapy with a small test dose or not. *Roenigk* et al. [84] suggest a test dose of MTX consisting of two 2.5-mg tablets (total 5 mg) is given and the CBC and platelet count are checked the next day. One week later the full dose of 15–25 mg of MTX is given orally. Among my own 248 psoriatics started on a dose of 20–25 mg MTX I have only observed 3 patients with a mild microcytic anemia [66], while 2 patients had a low platelet count and one had a nonfatal leukopenia during the first month after initiation of MTX therapy. A test dose of 5 mg MTX might give a false feeling of safety as it may not be enough to trigger a bone marrow depression or this latter develops later, i.e. after 2–4 weeks with 20–25 mg MTX.

Drug Dosage Schedules of MTX Therapy in Psoriasis

Since 1948 MTX has been utilized in the treatment of leukemia, in particular acute lymphatic leukemia, in the attempt to inhibit the increased newformation of cancer cells in the patients. It was of interest, therefore, to administer MTX at short intervals in order to inhibit the newformation of cancer cells in the phase of synthesis. This also caused inhibition of cell formation in normal tissue as a side effect. Accordingly, the administration of MTX had to be adapted to obtaining the greatest possible effect with the fewest possible side effects. In cancer therapy considerable side effects could be accepted, but in benign diseases, such as psoriasis, they might result in MTX therapy being restricted or even completely abandoned.

In 1950, *Ferguson* et al. [28], in animal experiments, found that MTX was less toxic (about 1/10) per weight unit than aminopterin and that a single, high dose of MTX was less toxic than several low doses of the same total quantity. This last observation was confirmed by *Rall* et al. [47] giving dogs MTX parenterally in doses from 0.06 to 100 mg/kg.

In 1960, *Condit* [17] found surprisingly little toxicity of large (up to 1,000 mg), infrequent MTX doses in 10 cancer patients. In 1966, *Bruce* et al. [11] reported from studies on mice that the toxicity of folic acid antagonists was directly proportional to the duration of exposure to the drug rather than the concentration of the drug in the body.

In 1971, *Zaharko* et al. [113] predicted, on the basis of pharmacokinetic studies in man, that the divided doses of MTX might be expected to be more toxic than a single, large dose (1 mg/kg intravenously in 2 persons).

As for psoriasis, *Weinstein and Frost* [107] nevertheless recommended administering the weekly oral dose of MTX in three portions at 12-hour intervals. By thus dividing the dose, and prolonging the duration of treatment and thereby the presence of the drug in the organism, they hoped to prolong the time during which the basal layer with its increased newformation of cells in psoriasis was exposed to MTX. From the point of view of cell kinetics in the epidermis this was a good idea, but from the pharmacokinetic and hepatotoxic point of view this therapeutic principle appears unfortunate, because it entails longer-lasting exposure to MTX of all cells in the organism – and thus also more side effects.

Dahl et al. [21] published a material of 44 MTX-treated psoriatics among whom hepatic cirrhosis and fibrosis proved more common following daily oral doses than weekly oral doses of MTX. The doses varied from 2.5 mg daily to 5–45 mg intramuscularly at intervals of 1–4 weeks.

Hepatic cirrhosis and fibrosis were more common in psoriatic patients who had received oral MTX more than 12 times a month than among those treated with fewer monthly doses (33 patients) [73]. *Roenigk* et al. [83] reported on two series of liver biopsies, one from psoriatics treated with three weekly oral doses (5.0–7.5 mg) of MTX and another one from psoriatics receiving one weekly oral dose of 25 mg MTX. Among the patients treated thrice weekly they found fibrosis to be more common ($p = 0.092$) than among those given one weekly dose (43 patients). The original principle of daily administration of MTX was abandoned in most dermatological clinics in the early 1970s. The single weekly oral dose is the easiest for the patient to remember.

In the MTX guidelines [84], the above is expressed as follows: There are two commonly used dose schedules of MTX for psoriasis. (1) single weekly oral, intravenous, or intramuscular intermittent large dose (25–35 mg), and (2) three divided-dose intermittent oral schedules over a 36-hour period each week.

Based on published studies, relatively equal effectiveness occurs with both schedules.

Studies to date indicate that when a certain amount of drug is given in one dose, there are less toxic effects than when the same amount is dispensed in divided doses over an expanded time period of 5–7 days [28, 73, 74, 83, 112, 113].

Specific Dose Schedule Recommendation

Dosage schedules cited below pertain to the average 70 kg adult. An initial test dose of 5–10 mg should be used to detect any idiosyncratic reaction.

(1) Weekly single oral, intramuscular, or intravenous dose schedule: (a) The range of oral dose schedule ordinarily is 7.5 to 25 mg/week, with occasional patients needing a maximum of 37.5 mg/week. It is recommended to be gradually increased (2.5 to 5 mg/week) with appropriate monitoring of the blood counts. (b) The range of intramuscular or rapid ('push') intravenous dosage schedule ordinarily is 7.5–50 mg/week, with occasional patients needing a maximum of 100 mg/week. Slow intravenous drips of MTX should never be used [11]! As the total dose is gradually increased, blood counts should be more carefully monitored.

(2) Divided oral dose schedule: 2.5–5.0 mg at 12-hour intervals for three doses each week. Dosage may be gradually increased by 2.5 mg/week, but the total dose should ordinarily not exceed 25 mg/week with appropriate monotoring of the blood counts.

All schedules should be continually tailored to the individual patients. Patients on any schedule should have dosage reduced and rest periods increased to the greatest possible extent.

The goal is not necessarily clearing of psoriasis, but rather to achieve adequate control at the lowest possible dosage and the longest rest period off MTX. The use of MTX may permit the return to conventional topical therapy; this course should be encouraged.

Continuing Laboratory Studies during MTX Therapy

It is important that the control laboratory tests are taken on the optimal day after the intake of MTX. A bone marrow depression, i.e. the maximum depression of the leukocyte and platelet counts, is most likely to be detected between days 5 and 7 after the last MTX dose. I have taken my laboratory control tests either on day 5 or 6 after MTX intake, particularly the first weeks after initiation of MTX therapy in which time period severe bone marrow depressions may occur. A significant decrease in leukocyte or platelet counts may necessitate reduction (50%, 75%) or temporary discontinuation of MTX therapy. Leukopenia continuing beyond a week deserves special hematologic elucidation. In the review on hepatotixicity of MTX [69] in psoriasis, I shall discuss the use of liver enzymes as control liver function tests. They should be taken a week or more after last MTX dose.

As shown by *Hendel and Nyfors* [38] long-term MTX therapy is a potential for folate depletion. Regular control of erythrocyte folate concentration with reference to a pre-MTX value is therefore recommended whenever long-term MTX therapy is initiated. The MTX guidelines [84] schematizise the control laboratory studies as follows: (1) 1- to 4-week intervals: leukocyte count and platelet count. (2) 3- to 4-month intervals: hemoglobin, urinalysis, serum creatinine, SGPT, SGOT, and alkaline phosphatase. (3) Yearly intervals: creatinine clearance (optional) and chest X-ray.

So far the guidelines, but the control laboratory studies must always be selected after the symptoms and findings of the patient in MTX therapy (see section on adverse effects).

Efficacy of MTX Therapy in Psoriasis

Edmundson and Guy [23] found more than 50% improvement of skin lesions in three-quarters of 17 psoriatics treated with MTX. These patients received 2.5 mg MTX daily by mouth for 6 days, followed by a withdrawal for 3 days, and then another 6-day course of MTX with a total of 30 mg in 12 doses. Repetition of a course of 12 doses was not prescribed until three or more months had elapsed after the original therapy. *Roenigk* et al. [81] observed complete clearing in 21 out of 45 psoriatics (47%) treated for an average of 15 weeks with one weekly oral dose up to 25 mg MTX. *Callaway* et al. [12] also reported complete clearing in 22 of 51 (43%) psoriatics and excellent improvement in another 18 after an average treatment period of 3–4 weeks, using one weekly oral dose up to 50 mg. In most patients satisfactory clinical results were achieved within 3–4 weeks after treatment was started. In 1969, *Roenigk* et al. [82] found complete clearing of psoriasis in 61% of 204 psoriatics treated by mouth once weekly with doses of 25 mg, in a few resistant cases 37.5 mg. The mean length of follow-up studies was 16 months (range: 2–50 months). In 1970, *Nyfors and Brodthagen* [63] observed complete clearing in 31 of their 50 (62%) MTX-treated psoriatics followed for a minimum of 1 year. Another 10 of the patients had more than 50% clearing and 6 patients (12%) less than 50% clearing, while 3 patients (6%) observed no effect of MTX therapy. The treatment consisted in one weekly oral dose of MTX up to 25 mg. After the same dosage *Nyfors* [65] found complete clearing in 52% of 248 psoriatics who had been treated for an average of 37 months (range: 2–105 months). In the remaining 120 patients with severe recalcitrant psoriasis was found almost clearing in 94 patients (38%) and some clearing in 21 patients (9%), while 5 patients (2%) did not benefit from MTX therapy.

In summary, MTX therapy in weekly oral doses of up to 25 mg for severe recalcitrant psoriasis is clearing at least 50% of patients and is very beneficial to about 80–90% of the patients.

Efficacy of MTX Therapy in Psoriatic Arthritis

Little has been published concerning the effect of MTX upon psoriatic arthritis. In 21 MTX-treated patients with psoriatic arthritis, *Black* et al. [8] found MTX to be better (decreasing joint tenderness and swelling, improving joint range of motion, and decreasing erythrocyte sedimentation rate) than placebo ($p<0.01$). In my own material [65] of 41 MTX-treated patients with psoriatic arthritis, 15 (37%) were relieved of pain, and in another 15 (37%) the pain had been appreciably mitigated, while in the remaining patients the MTX therapy had no effect upon the articular pain, possibly because the treatment had been instituted after destructive changes had developed in the joints. In 52 psoriatics with arthritis, *Kragballe* et al. [49] found a modest improvement in 21 (40%), while in the remaining 22 patients (42%) the signs of inflammatory joint disease almost disappeared. The patients had been treated for a median of 3 years. The favorable response was related to the short duration of disease. In 9 of the 52 patients (17%) MTX was withdrawn (after sufficient clinical improvement) without recurrence of arthritis.

In summary, MTX therapy in weekly oral doses up to 25 mg has been very beneficial in ⅓ to ½ of patients with psoriatic arthritis. Treatment should start early in the arthritis.

Effects of MTX Therapy in Generalized Pustular Psoriasis

In 1969, *Ryan and Baker* [88] published their analysis of effect of MTX therapy in generalized pustular psoriasis. They observed (1) clearing of pustules in 37 out of 47 patients (79%); (2) least effect of MTX in patients previously treated with systemic corticosteroids, and (3) that MTX therapy had made it possible to level systemic corticosteroids. In the 1980s retinoids have been a very good alternative in the treatment of pustular psoriasis [103, 105].

Rebound on Withdrawing MTX

A serious relapse of psoriasis may occur on withdrawing systemic corticosteroid treatment. In some cases this is followed by erythroderma or generalized pustular psoriasis [88]. Similarly, erythroderma may be observed after discontinuation of MTX therapy, which may be necessary, for instance, after liver fibrosis/cirrhosis has been diagnosed [6]. Neither erythroderma nor generalized pustular psoriasis occurred after discontinuation of MTX in my material [66]. On the other hand, discontinuation of MTX was followed in most cases by relapse of psoriasis of varying severity and after a varying length of time. In psoriatic arthritis *Kragballe* et al. [49] – as mentioned above – were able to withdraw MTX without recurrence in 9 of 52 patients (17%).

Effectiveness of Further Courses of MTX

In my experience, some psoriatics respond less well to MTX after having been treated for years. This is illustrated by the decrease in the percentage of complete clearing from our first analysis in 1970 [63] to the most recent analysis in 1978 [65], viz. from 62 to 52.

Precautions

MTX should be used cautiously and for restricted indications [65]. It has a potential for severe toxicity [62] and precautions should be taken. However, with careful management, acute toxic effects should seldom occur. This form of therapy should be directed only by physicians experienced with the drug, i.e. those who are familiar with its metabolic pathways, absorption, excretion and toxic manifestations.

Adverse Drug Experience Related to MTX Therapy

The nomenclature varies considerably as regards unwanted effects of drugs. Terms like adverse reactions, toxic effect, and side effects are often used synonymously, but the most common term seems to be side effects.

Definitions

Side effects are taken to mean harmful, unwanted effects of a drug in doses normally used for human subjects for profylatic, diagnostic, or therapeutic purposes.

A number of the unwanted effects of MTX may be designated side effects, meaning: unwanted effects which are a direct consequence of the known pharmacological properties of the drug.

Others may be called adverse reactions, meaning: unwanted effects of a drug due predominantly to unknown properties of the drug, not directly related to its desired pharmacological actions.

In practice, it is not always possible to distinguish between these two types of unwanted effects. In the following survey, therefore, the classification will be based upon the target organ as far as possible.

Symptoms

In a clinical study it is natural to divide side effects into acute, subacute and chronic and into those which cause symptoms and those which do not [27]. The subjective symptoms following MTX therapy may be: nausea, fatigue, oral discomfort and pain, indigestion, headache, dizziness, a burning sensation of the skin, decreased appetite, reduced libido, impaired memory and so forth. Among nonsymptomatic side effects there are, for example, milder degrees of hepatic fibrosis and cirrhosis.

Callaway et al. [12] pointed out the 2 most common subjective symptoms following MTX intake. After single oral doses of maximum 50 mg MTX weekly, they found nausea in 10–20% and oral discomfort in 3 (6%) of 51 psoriatics. *Roenigk* et al. [82] found nausea in 31%, fatigue in 7%, oral discomfort in 6%, and headache in 5% of 204 psoriatics treated with one weekly oral dose of maximum 37.5 mg MTX. In 1970, *Nyfors and Brodthagen* [63], studying 50 psoriatics treated with single weekly oral doses of maximum 25 mg MTX, found nausea in 12 (24%), fatigue in 8 (16%), and oral discomfort in 6 patients (12%). A transient burning sensation in the skin was experienced by 3 of 27 patients (11%) [104] and by 3 of 45 psoriatics (7%) on MTX therapy [81]. In my own material of 248 psoriatics who received one weekly oral dose of maximum 25 mg, the frequency of subjective symptoms was as follows: nausea 52%, fatigue 26%, oral discomfort 23%, abdominal pain 14%, and headache 11% [65].

In all the above-mentioned studies nausea was the most common symptom. In many patients it has been found to depend upon the dose of MTX. Most of my patients with nausea after a weekly dose of 15–25 mg reported that they obtained relief after the dose was reduced to below 15 mg.

It is difficult to measure the intensity of a symptom such as nausea. Often, this ends in a binary classification into: does not prevent the patient from working/prevents the patient from working. In my material of 248 psoriatics [66], the MTX therapy was temporarily withdrawn in 37 patients because of the following symptoms: nausea 30, impaired power of concentration 2, abdominal pain 2, reduced libido 1, repeated biliary colics 1, and headache 1. Another dimension of a symptom which is of great importance to the patient is its duration. As a rule, the nausea arose from 1–2 to 8–12 h after the intake of MTX and lasted for 1–3 h up to 1–2 days. A few patients were nauseated for 4–7 days after taking MTX. Constant nausea, even after reduction of the MTX dose, usually lead to rapid discontinuation of MTX medication. In the above-mentioned material of 248 patients, MTX was withdrawn because of nausea in about one-fifth of patients with this complaint.

'Objective symptoms' refers to the patient's own observations, such as vomiting, hair loss and erythema. They will be described under the discussion of organ-related side effects.

Side Effects from Different Organs

Bone Marrow

Leukopenia. Bone marrow suppression with fatal outcome has occurred after MTX treatment for psoriasis with a maximum dose of 25 mg weekly. *Black* et al. [8] have reported 1 case, *Nyfors* [62] 1 case, and *Ryan and Baker* [88] 3 cases. The last 4 deaths occurred after MTX treatment of generalised pustular psoriasis initially treated systemically with glucocorticoids. *Baker* [5] reviewed 18 deaths after MTX treatment reported at that time. He found 8 deaths due to bone marrow hypoplasia and regarded 12 of 18 deaths as being due to overdosage of MTX. In 1977, *Baker* [6] reviewed 38 deaths among MTX-treated psoriatics in the London area. He ascribed these deaths to corticosteroid complications 11, unrelated causes 10, infection 5, malignancies 5, cirrhosis complications 4, acute hepatorenal failure 2, and MTX withdrawal 2. *Roenigk* et al. [82] described 2 deaths, from septicemia with pneumonia and from pancytopenia, respectively, among 204 MTX-treated psoriatics. In my own material of 248 MTX-treated psoriatics there was 1 case of nonfatal leukopenia [65].

Low Platelet Counts. Trombocytpenia has been reported in MTX-treated psoriatics by several authors. *Black* et al. [8] found less than 100,000 platelets/ml in 1 of 21 MTX-treated psoriatics with psoriatic arthritis (5%), *van Scott* et al. [104] observed trombocytopenia in 1 of 27 (3.7%), *Roenigk* et al. [82] in 5 of 204 patients (2.5%), and *Moschella* [59] in 2 of 246 patients (0.8%). In 1 patient with pancytopenia, massive gastrointestinal hemorrhage was seen at autopsy [82]. In my own patients, MTX had to be interrupted temporarily in 2 out of 248 patients because of low platelet counts. A decreased number of platelets occurred most often in patients with hepatic cirrhosis [65].

Anemia. Anemia has arisen in MTX-treated psoriatics according to *Borrie and Clark* [10] and *Roenigk* et al. [82]. *Borrie and Clark* [10] reported 1 case of megaloblastic anemia, while *Roenigk* et al. [82] found anemia in 4 of 204 patients (2%). All the patients with anemia recovered. Among my own 248 patients I have observed only 3 cases of mild microcytic anemia [65].

Development of Leukemia-Like State. A few cases of leukemia-like/lymphoma-like state in MTX-treated psoriatics are on record [8, 33, 47, 92]. The development of such a leukemia-like state seems to be explicable – according to *Gailani* [33] – by bone marrow overshot from MTX-suppressed bone marrow following discontinuation of the drug. None of the 248 MTX-treated psoriatics [65] developed such a clinical picture, but a follow-up study in 1982 revealed that one had developed malignant lymphoma and one acute myloid leukemia during therapy [68].

Liver

For liver side effects, the reader is referred to the articles by *Nyfors* [69] and *Zachariae* [112].

Lungs

In 1969, *Clarysse* et al. [16] reported the first 7 cases of pulmonary infiltrations in MTX-treated patients with acute lymohoblastic leukemia.

The first case of 'MTX pneumonitis' in a psoriatic was reported in 1971 by *Filip* et al. [29]. In 1975, another case was published by *From* [32]. The clinical signs of a 'MTX pneumonitis' were a poor general condition with fever, cough, dyspnea, and cyanosis. Auscultation revealed only mild and uncharacteristic findings. X-rays showed basally widespread, bilateral infiltrations. Differential white cell count disclosed up to 20–25% eosinophils. The authors concluded that the condition may necessitate systemic administration of corticosteroids. In *Clarysse* et al.'s [16] material 3 out of 7 patients died. One patient relapsed when MTX therapy was resumed. This pulmonary condition seems to be explicable as a drug reaction to MTX, and the authors found that a viral infection in explanation of the condition has not been demonstrated. *Roenigk* et al. [82] reported 1 death from pneumonia and septicemia among their 204 MTX-treated psoriatics. In my own study [65] transient bilateral basal lung infiltrates were seen in 2 patients and transient unilateral basal lung infiltrates in 4.

Genital Tract

Actions upon the sperm during MTX therapy of psoriatics were reviewed in a paper by *Grunnet* et al. [35]. MTX therapy did not appear to affect the quality of the sperm in psoriatics. In specimens from 10 MTX-treated psoriatics normal sperm was found significantly more often than in specimens from 10 non-MTX-treated psoriatics. The frequency of changes in semen during MTX treatment cannot be given as most clinical studies have comprised only about 10 patients and have often included only post-MTX semen samples. In 1980, *Sussman and Leonard* [101] published a case report of a young man who twice developed severe (but reversible) oligospermia during MTX therapy.

Kidneys

In cancer therapy high doses of MTX (3 mg/kg) have been known to entail widespread epithelial necroses in the convoluted tubules [18].

During MTX treatment of psoriatics (0.3–0.4 mg MTX/kg/week) there are no records of renal damage as judged by the creatinine clearance, microscopic examination of the urine, study of the urine for blood and urea [6, 82]. Renal biopsies from 6 MTX-treated psoriatics have shown normal renal histology [110]. In my own material [65] there was no renal damage, assessed by the creatinine clearance, that could be attributed to MTX therapy. *Hendel and Nyfors* [37] studied the renal elimination of MTX in psoriatics and found a nonlinear elimination kinetics of MTX due to saturation of renal tubular reabsorpton.

Bladder Epithelium

There seems to be no reports on the influence of MTX upon the bladder epithelium in MTX-treated psoriatics. In my own material [65], cytological follow-up studies of urine specimens showed no atypical cells.

Gastrointestinal Tract

Among clinical gastrointestinal side effects in MTX-treated psoriatics, the following may be mentioned: oral ulcerations, vomiting, and duodenal ulcer. Oral ulcerations were found by *Hunter* [40] in 4 of 119 patients (3%), by *Callaway* et al. [12] in 3 of 51 patients (6%), by *Roenigk* et al. [82] by 12 of 204 patients (8%), and by *Moschella* [59] in 15 of 246 patients (6%). Vomiting was reported in 12 of 204 patients (6%) by *Roenigk* et al. [82] and in 10 of 246 patients (4%) by *Moschella* [59]. *Shrank and Blendis* [96] in 1965 presented the case report of a MTX-treated psoriatic having a large melena stool. Examination showed duodenal ulcer. The patient had had a duodenal ulcer diagnosed also in 1952. Other authors have reported gastrointestinal bleeding in MTX-treated psoriatics. *Black* et al. [8] reported gastrointestinal hemorrhage at autopsy in 1 of 21 patients, and *Roenigk* et al. [82] found massive gastrointestinal hemorrhage in 1 of 204 patients. My own material [65] includes several (< 10) patients in whom a previous, radiologically confirmed, peptic ulcer recurred during MTX therapy. Melena during MTX therapy was observed in 1 of these patients. Besides, 2 patients had recurrence of gallbladder colics [65].

Influence of MTX upon Hair

MTX in doses of 10–25 mg weekly affects the growth of hair. MTX affects the hair only in the anagen phase, also called the growth phase, during which Crounse and van Scott (20) found a reduced diameter of the hair bulbs with subsequent narrowing of the shafts. This narrowing occurred each time a dose of MTX (2.5–1,000 mg) had been administered. Diffuse hair loss in MTX-treated psoriatics has been reported by *O'Brien* et al. [70] in 1 of 18 patients (5.5%), by *Black* et al. [8] in 1 of 21 patients (5%), by *Roenigk* et al. [82] in 12 of 204 patients (6%) and by *Moschella* [59] in 12 of 246 patients (4.8%). In my own material (65) 6% of patients complained of diffuse hair loss.

Epidermis

Ulceration of the skin in MTX-treated psoriatics has been reported by *Black* et al. [8], with bleeding in 4 of 18 patients (19%), by *Rees* et al. [79] as ulceration in 1 of 168 patients (0.6%), by *Roenigk* et al. [82] in 6 of 204 patients (3%) and by *Moschella* [59] in 1.6% of 246 patients. *Van Scott* et al. [104] found erythema of the skin in 3 of 27 MTX-treated psoriatics (11%). In my own material [65], I have not observed ulcerations like those described by other authors. During my 7 years in Norway I have only seen one MTX-treated psoriatic (out of about 50 patients) with such an ulceration on the leg. Recently, *Lawrence and Dahl* [50] reported 6 patients with necrosis and ulceration associated with MTX therapy for psoriasis in doses of 10–20 mg weekly. *Reed and Sober* [75] described in 1983 a toxic epidermal necrolysis-like picture in a woman after 7.5 mg MTX for psoriasis.

Nervous System

In psoriatics there have been no reports – to the best of my knowledge – of neurotoxicity after MTX therapy. The explanation may be that MTX does not pass the blood-brain barrier to any major extent. On the other hand, dementia may occur after folic acid deficiency which may affect psoriatics after long time on MTX [100]. *Roenigk* et al. [81] reported that convulsions had occurred in 1 of 45 patients (2%). In my own material [65] 2 patients complained of impairment of memory and concentration ability after the weekly administration of MTX. This was their sole complaint, but it was of a severity which made them wish to discontinue the MTX therapy for a time. These complaints were objectively reversible in both patients.

Other Clinical Adverse Effects

Among other clinical findings in MTX psoriatics it is worth mentioning: phlebitis in 4 patients (1.6%) and photosensitivity in 8 of 246 patients [59] and in 10 of 204 patients (5%) [82]. *Korossy and Hood* [48] described MTX reactivation of a sunburn reaction in a woman treated with daily MTX doses of 2.5 mg every other week for a choriocarcinoma. In 1981, *LeVine* [52] described a tender diffuse erythema involving all exposed areas developed 24 h after a treatment with psoralens and long-wave ultraviolet light (PUVA 3 J/cm^2) and MTX (25 mg intramuscularly) coincided. In 1982, *Armstrong and Poh-Fitzpatrick* [3] reported reactivation of a sunburn in a patient who was receiving both coal tar with ultraviolet radiation for cutaneous lesions and 25 mg of MTX for psoriatic arthritis. *Chahidi* et al. [13] surveyed photosentization by MTX photoproducts in 1983. Care must be used when using UV radiation and MTX in combination as reported early in 1969 by *Møller* [60]. In my own study [65], where MTX was used alone, I saw no photoeffect of MTX in doses up to 25 mg weekly.

Osteopathy has also been reported after MTX therapy [93].

Overdosage

The MTX guidelines [84] advise the following:

Leucovorin calcium (citrovorum factor) is an antidote for MTX. Where overdosage of MTX is suspected, the dose of leucovorin should be equal to or higher than the dose of MTX, and should be administered parenterally as soon as possible, within 12–24 h. Beyond 24 h after the last dose of MTX, there is little if any likehood that leucovorin will have any effect except in patients with poor drug excretion. Leucovorin may be administered by intravenous infusions in doses of up to 75 mg within 12 h, followed by 12 mg intramuscularly every 6 h for four doses. Leucovorin may be given intramuscularly (6–12 mg) every 6 h for four doses. In case of dehydration or renal impairment, leucovorin should be continued until it is certain that MTX has been excreted. Leucovorin should be readily available, since overdosage of MTX is an emergency situation and time waiting to get leucovorin from the pharmacy may be valuable time. So far the guidelines. *Zachariae and Bjerring* [111] have succesfully used 100 mg of MTX intravenously with an addition of leucovorin to clear 3 patients with psoriatic erythroderma. *Kavli and Volden* [46] have used leucovorin 24 h after MTX with good results in a 35-year-old female with pustular psoriasis who developed trombocytopenia (110,000) during MTX treatment. The use of leucovorin or citrovorum factor has been reviewed by *Sadee* [90].

Carcinogenicity

A development of malignant diseases during the MTX therapy for psoriasis has been feared because of the immunosuppresive effect of MTX. Apart from case reports, there seems to have been only 2 major studies of the development of internal cancers during MTX therapy for psoriasis. *Bailin* et al. [4] followed 205 MTX-treated psoriatics, but observed no increase in the incidence of internal cancers, neither on the whole nor of any special type of cancer. The follow-up period was a minimum of 7 years.

At a follow-up examination (median follow-up: 7 years) in 1982 of my own material of 248 MTX-treated psoriatics, the following malignancies were diagnosed during MTX therapy: ovarian cancer 3, breast cancer 2, esophageal cancer in 1 male, pancreatic cancer in 1 female and 1 male, and 1 squamous cell carcinoma of the scrotal skin. As the observed number of malignant neoplasms was considerably smaller than the expected number, it was concluded that MTX therapy as used in the treatment of psoriasis does not seem to contribute to the development of malignant neoplasms [68]. *Stern* et al. [98] came to the same results. This and related questions have been discussed by others [34, 54].

Chromosomal Changes

Ryan et al. [87] studied chromosomes in cultured lymphocytes from peripheral blood of 8 MTX-treated psoriatics and 8 controls with psoriasis. Chromosome gaps and breaks (where both chromatids are affected at the same level), or separated fragments of chromosomes were seen in 10% of cells in the 8 treated patients and in 4% in the controls. The authors concluded that MTX was potentially harmful in the long run. *Jensen and Nyfors* [41] in a chromosome study of post-MTX specimens of the bone marrow from 11 psoriatics, found a statistical significant ($p < 0.01$) increase in the number of chromatid breaks and abnormal chromosome types compared to 10 normal controls. On the other hand, there was no significant increase in chromosomal abnormalities in lymphocyte cultures from 9 of the MTX-treated psoriatics, so that the bone marrow method seems better suited for detecting chromosomal abnormalities. *Nielsen and Zachariae* [61], comparing 10 patients having severe psoriasis with 41 controls, found an increased frequency ($p < 0.05$) of chromosomal gaps, chromosomal breaks, dicentric chromosomes, and acentric chromosomal fragments in leukocytes cultured for 48 h. *Jensen and Nyfors* [42], comparing cytogenetic studies on bone marrow and peripheral blood from 22 MTX-treated psoriatics with the results obtained with the micronucleus test on bone marrow smears from 25 MTX-treated psoriatics, found the micronucleus test was more sensitive than the chromosome analysis as regards the clastogenic effect of MTX. In the bone marrow preparations an increased number of structural chromosomal aberrations was present, whereas abnormalities were not observed in the peripheral blood cells. An examination of bone marrow chromosomes must therefore be included in the study of the possible chromosome-breaking effect of chemical agents.

Embryonic Damage

In the 1960s MTX was used as an abortifacient. *Milunsky* et al. [55] reported the case of a newborn with a large number of chromosomal abnormalities and an abnormal number of digits born to a mother who had taken MTX in daily oral doses of 2.5 mg for 5 days. In my own study [65] no women conceived during MTX therapy, but 3 men on MTX made their wives pregnant. One of the wives wanted and had an abortion, while the others gave birth to normal infants. Since MTX is a known teratogenic agent and abortificiant [1] and may cause reversible oligospermia in men [101], besides that it may accumulate in different organs for up to 4 months [15], both male and female patients planning to conceive children should be off MTX 6 months prior to conception. A similar view has been given by *Perry* [72].

Combination Therapy for Psoriasis

Combination of different treatment modalities [85] is always temptating in order to increase efficacy of therapy and to reduce side effects, particularly the long-term side effects. *Dahl-Nyfors and Brodthagen* [22] confirmed in 1976 the old experience that UV-B irradiation is beneficial to psoriasis. They found, that 65–93% (5% confidence limits) of psoriatics cleared after treatment with UV-B light thrice weekly for 3 weeks [22]. In 1982, *Paul* et al. [71] treated 26 patients with extensive psoriasis with a 3-week course of weekly oral MTX given on weekends followed by a combination of ultraviolet B therapy given Monday, Wednesday and Friday and MTX. This protocol achieved clearance of psoriasis in all 26 patients in a mean of 7 weeks with 12 exposures to UV-B and a mean total dose of 112 mg MTX. No MTX erythema recall was seen. This combination therapy may be worth trying in psoriatic arthritis. Since 1975, treatment of psoriasis with psoralens and irradiation with longwave ultraviolet light, UV-A, called PUVA therapy, gained wide use. The same group who made PUVA therapy for psoriasis so well known published, in 1984 [99], a 5.7-year retrospective study of 1,380 PUVA-treated psoriatics. They found a *50-fold* increase in the incidence of squamous cell carcinoma in the high-dose PUVA group as compared with the expected frequency in the general population. This group of researchers has combined MTX with PUVA therapy [58]. Thirty psoriatics were treated with a 3-week course of MTX followed by a combination of PUVA therapy and MTX. Clearance was seen in 28 of 30 patients after a mean of 6 weeks and with a mean total dose of 93 mg MTX and a final UV-A irradiation dose at clearance of 6 J/cm^2. The only significant adverse effect was prolonged phototoxicity in 8 patients (27%). Besides the MTX erythema recall, this treatment combination combines 2 hepatotoxic drugs which seems less attractive. Retinoids have come to stay in the treatment of skin diseases, as reviewed by *Ward* [105]. *Rosenbaum and Roenigk* [86] were able to obtain clearing in a patient with recalcitrant generalized pustular psoriasis with combination therapy of etretinate and MTX, but not when they used MTX or etretinate alone. *Wanderveen* et al. [103] were able to treat a recalcitrant case of widespread pustular psoriasis combining MTX and etretinate. When the MTX dose was tapered – to reach a controlled transition from MTX to etretinate therapy alone – the patient noted increased psoriatic arthritis pain. Finally, well-known topical therapy such as Ingram therapy (tar baths, UV irradiation and dithranol) has been combined with MTX. *Van de Kerkhof and Mali* [102] found, in 98 patients with difficult psoriasis so treated, that the period before readmission was extended from 5 months to more than 3 years.

Reviews

The following papers are reviews: [1, 7, 9, 13, 14, 19, 24, 34, 39, 43, 44, 56, 57, 64, 66, 67, 69, 80, 84, 85, 90, 91, 95, 105, 106, 108, 112].

References

1 Anderson, T.F.: Psoriasis. Med. Clins. N. Am. *66:* 769–794 (1982).

2 Angier, R.B.; Boothe, J.H.; Hutchings, B.L.; Mowat, J.H.; Semb, J.; Stokstad, E.L.R.; Subbarow, Y.; Waller, C.V.: The structure and synthesis of the liver L. casei factor. Science *103:* 667–669 (1946).

3 Armstrong, R.B.; Poh-Fitzpatrick, B.: Methotrexate and ultraviolet radiation. Archs Derm. *118:* 177–178 (1982).

4 Bailin, T.L.; Tindall, J.P.; Roenigk, H.H., Jr.; Hogan, M.D.: Is methotrexate therapy for psoriasis carcinogenic? A modified retrospective-prospective analysis. J. Am. med. Ass. *232:* 359–362 (1975).

5 Baker, H.: Some hazards of methotrexate treatment of psoriasis. Trans. St. Johns Hosp. derm. Soc. *56:* 111–116 (1970).

6 Baker, H.: Methotrexate. The conservative treatment for psoriasis; in Faber, Cox, Psoriasis. Proc. 2nd Int. Symp., pp. 235–242 (Yorke Medical Books, New York 1977).

7 Baker, H. & Wilkinson, D.S.: Psoriasis; in Rook, Wilkinson, Ebling, Textbook of dermatology, I–II; 3rd ed., pp. 1315–1367 (Blackwell, Oxford 1979).

8 Black, R.L.; O'Brien, W.M.; Scott, E.J. van; Auerbach, R.; Eisen, A.Z.; Bunim, J.J.: Methotrexate therapy in psoriatic arthritis. J. Am. med. Ass. *189:* 743–747 (1964).

9 Bleyer, W.A. : The clinical pharmacology of methotrexate. New applications of an old drug. Cancer *41:* 36–51 (1978).

10 Borrie, P.; Clark, P.A.: Megaloblastic anemia during methotrexate treatment of psoriasis. Br. med. J. *i:* 1339 (1966).

11 Bruce, W.R.; Meeker, B.E.; Valeriote, F.A.: Comparison of the sensitivity of normal hematopoietic and transplanted lymphoma colony-forming cells to chemotherapeutic agents administered in vivo. J. natn. Cancer Inst. *37:* 233–245 (1966).

12 Callaway, J.L.; McAfee, W.C.; Finlayson, G.R.: Management of psoriasis using methotrexate orally in a single dose. South. med. J. *59:* 424–426 (1966).

13 Chahidi, C.: Photosentization by methotrexate photoproducts. Photochem. Photobiol. *38:* 317–322 (1983).

14 Champion, C.H.: Psoriasis and its treatment. Br. med. J. *282:* 343–345 (1981).

15 Charache, S.; Condit, P.T.; Humphreys, S.R.: Studies on the folic acid vitamins. IV. The persistence of amethopterin in mammilian tissues. Cancer *13:* 236–240 (1960).

16 Clarysse, A.M.; Cathey, W.J.; Cartwright, G.E.; Wintrobe, M.M.: Pulmonary disease complicating intermittent therapy with methotrexate. J. Am. med. Ass. *209:* 1861–1864 (1969).

17 Condit, P.T.: Studies on the folic acid vitamins. II. The acute toxicity of amethopterin in man. Cancer *13:* 222–228 (1960).

18 Condit, P.T.; Chanes, R.E.; Joel, W.: Renal toxicity of methotrexate. Cancer *23:* 126–131 (1969).

19 Cram, D.L.: Psoriasis. Current advances in etiology and treatment. J. Am. Acad. Derm. *4:* 1–14 (1981).

20 Crounse, R.G.; Scott, E.J. van: Changes in scalp hair roofs as a measure of toxicity from cancer chemotherapeutic drugs. J. invest. Derm. *35:* 83–90 (1960).

21 Dahl, M.G.C.; Gregory, M.M.; Scheuer, P.J.: Methotrexate hepatotoxicity in psoriasis – comparison of different dose regimens. Br. med. J. *i:* 654–656 (1972).

22 Dahl-Nyfors, K.B.; Brodthagen, H.: Treatment of psoriasis with carbon-arc light baths and 8-methoxypsoralen. Ugeskr. Laeg. *138:* 2221–2224 (1976).

23 Edmundson, W.F.; Guy, W.B.: Treatment of psoriasis with folic acid antagonists. Archs Derm. *78:* 200–203 (1958).

24 Farber, E.M.; Cox, A.J.: Psoriasis. Proc. 3rd Int. Symp. (Grune & Stratton, New York 1982).

25 Farber, S.; Cutler, E.C.; Hawkins, J.W.; Harrison, J.H.; Peirce, E.C.; Lenz, G.G.: The action of pteroylglutamic conjugates on man. Science *106:* 619–621 (1947).

26 Farber, S.; Diamond, L.K.; Mercer, R.D.; Sylvester, R.F.; Wolff, J.A.: Temporary remissions in acute leukemia in children produced by folic acid antagonist, 4-amino-pteroyl-glutamic acid (aminopterin). New Engl. J. Med. *238:* 787–793 (1948).
27 Feinstein, A.R.: Clinical judgment, pp. 131–132 (Williams & Wilkins, Baltimore 1967).
28 Ferguson, F.C., Jr.; Thiersch, J.B.; Philips, F.S.: The action of 4-amino-N[10]-methyl-pteroylglutamic acid in mice, rats, and dogs. J. Pharmac. exp. Ther. *98:* 293–299 (1950).
29 Filip, D.J.; Logue, G.L.; Harle, T.S.; Farrar, W.H.: Pulmonary and hepatic complications of methotrexate therapy in psoriasis. J. Am. med. Ass. *216:* 881–882 (1971).
30 Franklin, A.L.; Stokstad, E.L.R.; Belt, M.; Jukes, T.H.: Biochemical experiments with a synthetic preparation having an action antagonistic to that of pteroylglutamic acid. J. biol. Chem. *169:* 427–436 (1947).
31 Franklin, A.L.; Belt, M.; Stokstad, E.L.R.; Jukes, T.H.: Biological studies with 4-amino-10-methylpteroylglutamic acid. J. biol. Chem. *177:* 621–629 (1949).
32 From, E.: Methotrexate pneumonitis in a psoriatic. Br. J. Derm. *93:* 107–110 (1975).
33 Gailani, S.D.: Pseudoleukemia during recovery from methotrexate toxicity in psoriasis. N.Y. State J. Med. *72:* 1054–1056 (1972).
34 Gelfant, S.: Psoriasis versus cancer. Adaptive versus iatrogenic human cell proliferative diseases. Int. Rev. Cytol. *81:* 145–162 (1983).
35 Grunnet, E.; Nyfors, A.; Brogaard-Hansen, K.: Studies on human semen in topical corticosteroid-treated and in methotrexate-treated psoriatics. Dermatologica *154:* 78–84 (1977).
36 Gubner, R.; August, S.; Ginsberg, V.: Therapeutic suppression of tissue reactivity. II. Effect of aminopterin in rheumatoid arthritis and psoriasis. Am. J. med. Sci. *221:* 176–182 (1951).
37 Hendel, J.; Nyfors, A.: Nonlinear neural elimination kinetics of methotrexate due to saturation of renal tubular reabsorption. Eur. J. clin. Pharmacol. *26:* 121–124 (1984).
38 Hendel, J.; Nyfors, A.: Impact of methotrexate therapy on the folate status of psoriatic patients. Clin. exp. Derm. *10:* 30–35 (1985).
39 Hoffmeister, R.T.: Methotrexate therapy in rheumatoid arthritis: 5 years experience. Am. J. Med. *75:* 69–73 (1983).
40 Hunter, G.A.: The use of methotrexate in the treatment of psoriasis. Australas. J. Derm. *6:* 248–253 (1962).
41 Jensen, M.K.; Nyfors, A.: The cytogenetic effect of folic acid antagonists on human cells in vivo; in Hejzlar, Semonsky, Masak, Adv. Antimicrobiol. Antineoplas. Chemother., pp. 351–352 (Urban & Schwarzenberg, Wien 1972).
42 Jensen, M.K.; Nyfors, A.: Cytogenetic effect of Methotrexate on human cells in vivo. Comparison between results obtained by chromosome studies on bone-marrow cells and blood lymphocytes and by the micronucleus test. Mutat. Res. *64:* 339–343 (1979).
43 Jolivet, J.; Cowan, K.H.; Curt, G.A.; Clendeninn, N.J.; Chabner, B.A.: The pharmacology and clinical use of methotrexate. New Engl. J. Med. *309:* 1094–1104 (1983).

44 Jukes, T.H.: The history of methotrexate. Cutis *21:*396–398 (1978).
45 Jürgens, H.; Göbel, V.: Pharmacology, toxicology and drug interactions of methotrexate. Rheumatology, vol. 9, pp. 6–19 (Karger, Basel 1986).
46 Kavli, G.; Volden, G.: Psoriasis pustulosa treated with Methotrexate and leucovorin. Acta derm.-vener., Stockh. *63:*368 (1983).
47 Kelleter, R.; Kuhn, D.; Schnyder, U.W.; Schrøter, R.: Lymforeticuläre Hyperplasie nach dreijähriger Methotrexatbehandlung einer Psoriasis. Dt. med. Wschr. *97:* 514–516 (1972).
48 Korossy, K.S.; Hood, A.F.: Methotrexate reactivation of sunburn reaction. Archs Derm. *117:*310–311 (1981).
49 Kragballe, K.; Zachariae, H.; Zachariae, E.: Methotrexate in psoriatic arthritis. Acta derm.-vener., Stockh. *63:* 165–167 (1983).
50 Lawrence, C.M.; Dahl, M.G.C.: Cutaneous necrosis associated with methotrexate treatment for psoriasis. Br. J. Derm. *107:*suppl. 22, p. 24 (1982).
51 Leuchtenberger, R.; Leuchtenberger, C.; Laszlo, D.; Lewisohn, R.: The influence of 'Folic acid' on spontaneous breast cancers in mice. Science *101:* 46 (1945).
52 LeVine, M.J.: Erythema resulting from suberythemogenic doses of ultraviolet radiation and methotrexate. Archs Derm. *117:*656–658 (1981).
53 Lomholt, G.: Psoriasis. Prevalence, spontaneous course, and genetics (GEC Gad, Copenhagen 1963).
54 MacKie, R.M.; Fitzsimons, C.P.: Risk of carcinogenicity in patients with psoriasis treated with methotrexate or PUVA singly or in combination. J. Am. Acad. Derm. *9:*467–469 (1983).
55 Milunsky, A.; Graef, J.W.; Gaynor, M.F.: Methotrexate-induced congenital malformations. With a review of the literature. J. Pediat. *72:*790–795 (1968).
56 Moll, J.M.H.: The clinical spectrum of psoriatic arthritis. Clin. Orthoped. *143:* 66–75 (1979).
57 Moll, J.M.H.: Seronegative arthropathies. J. R. Soc. Med. *76:*445–448 (1983).
58 Morison, W.L.; Momtaz, K.; Parrish, J.A.; Fitzpatrick, T.B.: Combined methotrexate-PUVA therapy in the treatment of psoriasis. J. Am. Acad. Derm. *6:* 46–51 (1982).
59 Moschella, S.L.: Chemotherapy of psoriasis: ten years of experience. Int. J. Derm. *15:*373–378 (1976).
60 Møller, H.: Reactivation of acute inflammation by methotrexate. J. invest. Derm. *52:*437–441 (1969).
61 Nielsen, J.; Zachariae, H.: Chromosome aberrations in severe psoriasis. Acta derm.-vener., Stockh. *53:* 192–194 (1973).
62 Nyfors, A.: Methotrexate in psoriasis. Lancet *ii:* 1251–1252 (1968).
63 Nyfors, A.; Brodthagen, H.: Methotrexate for psoriasis in weekly oral doses without any adjunctive therapy. Dermatologica *140:*345–355 (1970).
64 Nyfors, A.; Lemholt, K.: Psoriasis in children. A short review and a survey of 245 cases. Br. J. Derm. *92:*437–442 (1975).
65 Nyfors, A.: Benefits and adverse drug experiences during long-term methotrexate treatment of 248 psoriatics. Dan. med. Bull. *25:*208–211 (1978).
66 Nyfors, A.: Methotrexate therapy of psoriasis. Effects and side effects with particular reference to hepatic damage. Dan. med. Bull. *27:*74–96 (1980).

67 Nyfors, A.: Psoriasis in children. Characteristics, prognosis and therapy. A review. Acta derm.-vener., Stockh., suppl. *95*, pp. 47–53 (1981).

68 Nyfors, A.; Jensen, H.: Frequency of malignant neoplasms in 248 long-term Methotrexate-treated psoriatics. Dermatologica *167:* 260–261 (1983).

69 Nyfors, A.: Methotrexate hepatotoxicity in psoriasis and psoriatic arthritis. A review. Rheumatology, vol. 9, pp. 192–212 (Karger, Basel 1986).

70 O'Brien, W.M.; Scott, E.J. van; Black, R.L.; Eisen, A.Z.; Bunim, J.J.: Clinical trial of amethopterin (methotrexate) in psoriatic and rheumatoid arthritis (preliminary report). Arthritis Rheum. *5:* 312 (1962).

71 Paul, B.S.; Momtaz, K.; Stern, R.S., Arndt, K.A.; Parrish, J.A.: Combined Methotrexate-ultraviolet B therapy in the treatment of psoriasis. J. Am. Acad. Derm. *7:* 758–762 (1982).

72 Perry, W.H.: Methotrexate and teratogenesis. Archs Derm. *119:* 874 (1983).

73 Podurgiel, B.J.; McGill, D.B., Ludwig, J., Taylor, W.F.; Muller, S.A.: Liver injury associated with methotrexate therapy for psoriasis. Mayo Clin. Proc. *48:* 787–792 (1973).

74 Rall, D.; Mead, J.A.R.; Freireich, E.J.; Scott, E.J. van; Hertz, R.; Lipsett, M.B.: Folic acid antagonists: effect on the cell and the patient. Ann. intern. Med. *59:* 931–956 (1963).

75 Reed, K.M.; Sober, A.J.: Methotrexate-induced necrolysis. J. Am. Acad. Derm. *8:* 677–679 (1983).

76 Rees, R.B.; Bennett, J.H.; Bostick, W.L.: Aminopterin for psoriasis. Archs Derm. *72:* 133–143 (1955).

77 Rees, R.B.; Bennett, J.H.: Further observations on aminopterin for psoriasis. J. invest. Derm. *32:* 61–66 (1959).

78 Rees, R.B.; Bennett, J.H.: Methotrexate versus aminopterin for psoriasis. Archs Derm. *83:* 970–972 (1961).

79 Rees, R.B.; Bennett, J.H.; Maibach, H.I.; Arnold, H.L.: Methotrexate for psoriasis. Archs Derm. *95:* 2–10 (1967).

80 Rees, R.B.: Methotrexate: 20 years later; in Farber, Cox, Psoriasis. Proc. 2nd Int. Symp., pp. 231–234 (Yorke Medical Books, New York 1977).

81 Roenigk, H.H., Jr.; Haserick, J.R.; Curtis, G.H.: Methotrexate for psoriasis. A preliminary report. Cleve. Clin. Q. *32:* 211–215 (1965).

82 Roenigk, H.H., Jr.; Fowler-Bergfeld, W.; Curtis, G.H.: Methotrexate for psoriasis in weekly oral doses. Archs Derm. *99:* 86–93 (1969).

83 Roenigk H.H., Jr.; Auerbach, R.; Bergfeld, W.F.; Frost, P.; Maibach, H.I.; Tindall, J.; Weinstein, G.D.: A cooperative, prospective study of the effects of chemotherapy of psoriasis on liver biopsies; in Farber, Cox, Psoriasis. Proc. 2nd Int. Symp., pp. 243–248 (Yorke Medical Books, New York 1977).

84 Roenigk, H.H., Jr.; Auerbach, R.; Maibach, H.I.: Methotrexate guidelines – revised. J. Am. Acad. Derm. *6:* 145–155 (1982).

85 Roenigk, H.H., Jr.; Maibach, H.I.: Methotrexate. Semin. Derm. *2:* 231–237 (1983).

86 Rosenbaum, M.M.; Roenigk, H.H., Jr.: Treatment of generalized pustular psoriasis with etretinate (RO 10-9359) and methotrexate. J. Am. Acad. Derm. *10:* 357–361 (1984).

87 Ryan, T.J.; Boddington, M.M.; Spriggs, A.I.: Chromosomal abnormalities produced by folic acid antagonits. Br. J. Derm. *77:* 541–555 (1965).

88 Ryan, T.J.; Baker, H.: Systemic corticosteroids and folic acid antagonists in the treatment of generalized pustular psoriasis. Evaluation and prognosis based on the study of 104 cases. Br. J. Derm. *81:* 134–145 (1969).
89 Ryback, R.S.; Eckardt, M.J.; Felsher, B.; Rawlings, R.R.: Biochemical and hematologic correlates of alcoholism and liver disease. J. Am. med. Ass. *248:* 2261–2265 (1982).
90 Sadee, W.: Antineoplastic agents: high dose methotrexate and citrovorum factor rescue. Ther. Drug Monit. *2:* 177–190 (1980).
91 Schornagel, J.H.: The clinical pharmacology of methotrexate. Cancer Treat. Rev. *10:* 53–75 (1983).
92 Schröter, R.; Kelleter, R.; Kuhn, D.: Reticulosis as a long-term side-effect of Methotrexate therapy in a patient with generalized psoriasis and psoriatic arthritis. Dermatologica *143:* 131–136 (1971).
93 Schwartz, A.M.: Methotrexate osteopathy. Skeletal Radiol. *11:* 13–16 (1984).
94 Seeger, D.R.; Smith, J.M., Jr.; Hultquist, M.E.: Antagonist for pteroylglutamic acid. J. Am. chem. Soc. *69:* 2567 (1947).
95 Shen, D.D.; Azarnoff, D.L.: Clinical pharmacokinetics of Methotrexate. Clin. Pharmacokinet. *3:* 1–13 (1978).
96 Shrank, A.B.; Blendis, L.M.: Folic acid antagonists in treatment of psoriasis. Br. med. J. *ii:* 156 (1965).
97 Smith Jr., J.M.; Cosulich, D.B.; Hultquist, M.E.; Seeger, D.R.: The chemistry of certain pteroylglutamic acid antagonists. Trans. N.Y. Acad. Sci. *10:* 82–83 (1948).
98 Stern, R.S.; Zieler, S.; Parrish, J.A.: Methotrexate used for psoriasis and the risk for noncutaneous or cutaneous malignancy. Cancer *50:* 869–872 (1982).
99 Stern, R.S.; Laird, N.; Melski, J.; Parrish, J.A.; Fitzpatrick, T.B.; Bleich, H.L.: Cutaneous squamous cell carcinoma in patients treated with PUVA. New Engl. J. Med. *310:* 1156–1161 (1984).
100 Strachan, R.W.; Henderson, J.G.: Dementia and folate deficiency. Q. Jl. Med. *36:* 189–204 (1967).
101 Sussman, A.; Leonard, J.M.: Psoriasis, methotrexate and oligospermia. Archs Derm. *116:* 215–217 (1980).
102 van de Kerkhof, P.C.M.; Mali, J.W.H.: Methotrexate maintenance following Ingram therapy in difficult psoriasis. Br. J. Derm. *106:* 623–627 (1982).
103 Vanderveen, E.E.; Ellis, C.N.; Campbell, J.P.; Case, P.C.; Vorhees, J.J.: Methotrexate and etretinate as concurrent therapies in severe psoriasis. Archs Derm. *118:* 660–662 (1982).
104 van Scott, E.J.; Auerbach, R.; Weinstein, G.D.: Parenteral methotrexate in psoriasis. Archs Derm. *89:* 550–556 (1964).
105 Ward, A.; Brogden, R.N.; Heel, R.C.; Speight, T.M.; Avery, G.S.: Etretinate. A review of its pharmacological properties and therapeutic efficacy in psoriasis and other skin diseases. Drugs *26:* 9–43 (1983).
106 Watson, W.; Farber, E.M.: Psoriasis in childhood. Pediat. Clins. N. Am. *18:* 875–895 (1971).
107 Weinstein, G.D.; Frost, P.: Methotrexate for psoriasis. Archs Derm. *103:* 33–38 (1971).
108 Willkens, R.F.: Reappraisal of the use of methotrexate in rheumatic disease. Am. J. Med. *75:* 19–25 (1983).

109 Willkens, R.F.: Randomized, double-blind, placebo-controlled trial of low dose Methotrexate in psoriatic arthritis. Arthritis Rheum. *27:*376–381 (1984).
110 Zachariae, H.; Grunnet, E.; Søgaard, H.: Accidental kidney biopsies in psoriasis. Br. J. Derm. *94:*655–657 (1976).
111 Zachariae, H.; Bjerring, P.: Methotrexate in psoriasis with and without Leucovorin: effect of different dosage schedules on the acute liver toxicity. Acta derm.-vener., Stockh. *62:*446–448 (1982).
112 Zachariae, H.: Psoriasis and the liver; in Roenigk, Jr., Maibach, Psoriasis (Dekker, New York 1985).
113 Zaharko, D.S.; Dedrick, R.L.; Bischoff, K.B.; Longstreth, J.A.; Oliverio, V.T.: Methotrexate tissue distribution. Prediction by a mathematical model. J.natn. Cancer Inst. *46:*775–784 (1971).

Prof. Allan Nyfors, MD, Department of Dermatology,
Haukeland University Hospital, N-5016 Bergen (Norway)

Rheumatology, vol. 9, pp. 88–95 (Karger, Basel 1986)

Low-Dose Methotrexate Therapy in Psoriatic Arthritis

A. Dunky

Krankenhaus der Stadt Wien-Lainz, 2. Medizinische Abteilung mit Rheuma-Ambulatorium, Zentrum für Diagnostik und Therapie Rheumatischer Erkrankungen, Wien, Österreich

Methotrexate (MTX) was initially used as an oncologic agent. Its relative safety, as well as the high benefit-to-risk ratio, made it an acceptable and attractive agent for use in chemotherapy.

The first to recognize the potential use in psoriasis and rheumatoid arthritis were *Gubner* et al. [2]. It has been used extensively in the treatment of psoriasis, resistant to other forms of therapy. There were different types of therapeutic regimens.

The use of MTX naturally extended from psoriasis to psoriatic arthritis. One of the first studies to assess its efficacy in psoriatic arthritis was carried out by *Black* et al. [1]. The first summarizing data on the treatment of psoriatic arthritis (PA) with MTX were given by *Wright and Moll* [20] in 1976. They introduced a uniform scheme of an i.v. treatment and proved the better therapeutic breadth of the i.v. administration.

We have been treating PA with MTX since about 1967 [15]. We started with an oral therapeutic regimen, but as there were a lot of side effects, we turned to the i.v. administration. Since 1972 we have treated severe forms of PA on an i.v. therapeutic scale with MTX. We start with a low dose and enhance it slowly, by monitoring the cumulative MTX dose. In this way, we treated a total of about 264 patients. This present report concerns a prospective analysis of a 2-year period (1982–1984) of 31 consecutive patients suffering from PA who were treated with MTX in our department.

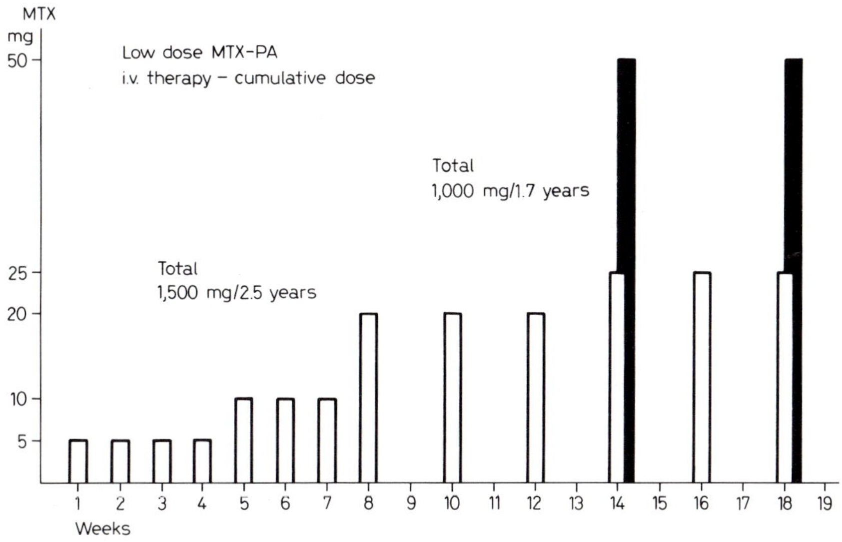

Fig. 1. MTX dosage was first started with 5 mg i.v. weekly, then increased to 10 mg/week, followed by 20 mg every second week and finally 50 mg/month. When a 'flair up' occurred with this dosage, then 25 mg MTX was administered every second week. For better control, the individual doses were added up to one total dose.

Material and Methods

31 patients were treated with MTX. All had classical seronegative psoriatic arthritis and manifest psoriasis. Only patients with PA not responding favorably to NSAD were considered as candidates for MTX treatment. Excluded from this treatment were patients who had severe renal and liver diseases or a hematological disorder. Women not using contraconceptives were also excluded from this basic treatment.

Before treatment, clinical and X-ray studies were made on the joints. All the patients were questioned about their age at the onset of their psoriasis vulgaris (PV) and also of the duration of their arthritic involvement.

Of the 31 patients, 22 were men and 9 were women. The age ranged from 30 to 79 (median age 52.7 years), weight 66 kg (51–86 kg), height 172.4 cm (156–184 cm). The duration of the PV manifestation ranged from 6 to 16 years (median duration 9.3 years). The median time period of the inflammatory joint manifestation was 3.3 years (with variations from 1 to 9 years).

The following skin manifestations were found with the patients included in this prospective study: 8 patients had skin manifestations on the scalp, 20 patients had PV manifestations on the trunk and/or the limbs (the percentage of the affected skin ranged from 6 to 37%). 3 patients had severe psoriasis on the nails. The joint manifestations were as follows; arthritis without psoriasis = 0; peripheral psoriatic arthritis = 21 (67%); psoriatic spondylitis with peripheral arthritis = 10 (33%). No patient had only spondylitic manifestations.

The analgesic consumption or the daily dose of nonsteroidal anti-inflammatory drugs in all patients was also established. The erythrocyte sedimentation rate (ESR) was determined and sera were analyzed by a latex fixation test. ANF and CRP were also analyzed. Before every MTX administration, a complete hematological and urine analysis was performed. Every 3 months a lab check (renal and liver function tests) was carried out.

At the beginning, MTX was administered weekly at an i.v. dose of 5 mg. When four doses had been administered and if no side effects had appeared, the weekly dose was raised to 10 mg i.v. After 3 injections the dose was raised to 20 mg i.v. every second week, for a further 3 administrations (fig. 1).

In patients with a good response, we continued this therapeutic regimen for a further 6 injections. Where the efficacy was poor, we raised the MTX administration to 50 mg i.v. every 4 weeks. The treatment was continued up to a cumulative dose of 1,000 mg MTX, which was reached after 1.7 years. The MTX treatment, the control of the cumulative dose and the patients' compliance were strictly observed. During treatment, joint involvement (pain on motion, tenderness and swelling), analgesic consumption and ESR were estimated.

The parameters were graded as: –1 (worsened), 0 (unchanged), 1 (improved), 2 (mild), and 3 (normalized).

By adding these scores, an improvement score was obtained for each patient.

Results

The influence of MTX on psoriatic arthritis was evaluated in all 31 patients in a maximal treatment period of 2 years. At the moment the patients are on a different level of a cumulative dosage of MTX.

22 male and 9 female patients with an average age of 52.7 years were observed. The average duration of arthritis involvement was 3.3 years. The average duration of psoriasis 9.3 years. 21 (67%) of these patients had peripheral joint involvement, 10 (33%) had psoriatic spondylitis including the sacro-iliacal joint and peripheral arthritis. Skin manifestation appeared in 8 patients only on the scalp. 3 had only nail psoriasis and 20 PV manifestations on the trunk and limbs.

The average duration of MTX treatment was 17.7 months by a median cumulative dose of 976.45 mg. In the following, the stage of the ongoing MTX treatment by cumulative dose regimen is described (fig. 2). There is no significant difference in intolerance between low and high cumulative doses. The highest range of MTX was 1,960 mg in 2 years. We do not usually go higher than 1,500 mg in 2.5 years. In our

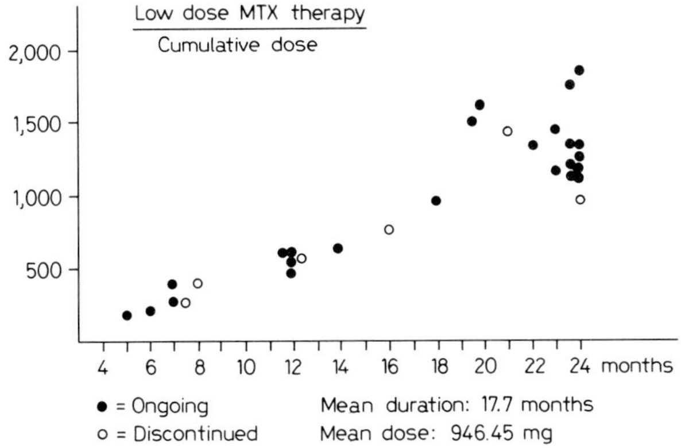

Fig. 2. Total dosage diagram in the current prospective study for psoriatic arthritis.

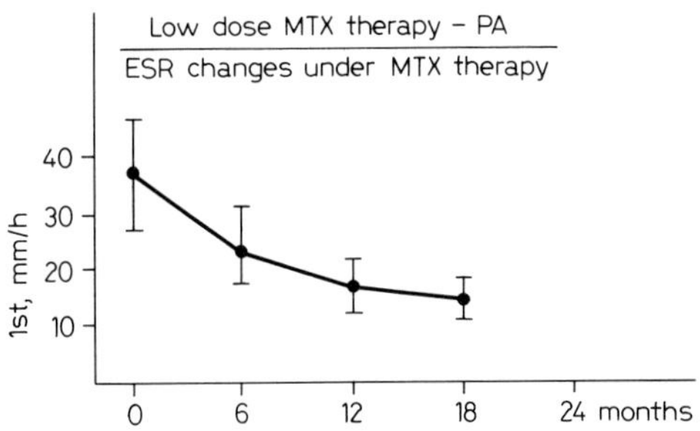

Fig. 3. As an expression of the disease-modulating effect and the antiphlogistic capacity, the sedimentation reaction decreased with MTX therapy.

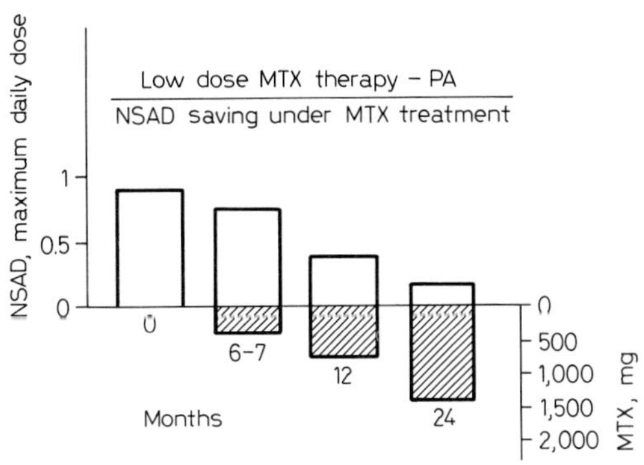

Fig. 4. Saving of the medication of nonsteroidal antirheumatic drugs with MTX therapy.

Table I. Low dose MTX therapy – PA: summary of clinical results

	n	%
Improvement	17	54.8
Indifferent	8	25.9
Discontinuation	6	19.3

earlier studies we saw a high incidence of side effects above a level of 1,500 mg within 2 years [15].

As seen in figures 3 and 4, there is a correlation between improvement of PA and duration as well as the cumulative dose of MTX treatment.

Therapy was discontinued in 6 patients (17.3%) for the following reasons: increased liver function test (we used no liver punction, but a control by scintigram [5, 6], 3 patients; disturbances of the hematopoetic system occurred in 2 patients, and 1 patient dropped out due to inefficacy after a dose of 450 mg MTX within 8 months.

After discontinuing MTX therapy, liver values improved within 4 weeks. At that time the total dose was at 250–820 mg MTX. These changes in the results of the blood tests were thrombocytopenia twice and leukopenia once. When those side effects appeared the patients had been administered 1,500 and 1,050 mg MTX , respectively.

In 17 patients, a distinct improvement of the clinical and laboratory parameters could be observed. A first improvement of the clinical parameter was observed between 150 and 220 mg MTX, whereas a significant improvement was observable only after a total dose of 300 mg MTX had been administered. At the time of the examination 8 patients (25.9%) showed no improvement compared to the initial values (table I).

Discussion

The purpose of the present study was to evaluate, for the future, the use of MTX in a larger group of patients suffering from psoriatic arthritis.

Our data argue that MTX may continue to play an important role in the medical management of the more severe forms of PA as a 'basic-treatment drug'. The results of this treatment seem to be more favorable the shorter the duration of the disease. MTX should be given with caution and monitored carefully. The best way to handle this is by 'cumulative dose monitoring', for the compliance of which a good cooperation of the patients is required.

By continuous controls of the hematopoetic system before every drug administration and with regularly carried out liver function tests, less side effects appear.

In long-term studies special attention should be given to liver toxicity, especially by a cumulative dosage of MTX of more than 1,500 mg in less than 1 year.

Also alcohol intake should be limited with patients on MTX. With i.v. administration we observe a satisfactory control and a good compliance by the patients.

In our study only 6 patients discontinued MTX treatment due to side effects. On the other hand, we saw a distinct improvement by 17 patients (54.8%). At the moment 8 patients seem to be in an indifferent condition. With these patients, duration of treatment is too short, and the level of the cumulation dose too low.

Our impression is that there is also a correlation between clinical/laboratory improvement and the dose/time ratio. Concomitant analgesic or anti-inflammatory drugs may play a role on plasma protein binding or renal excretion of MTX. On the other hand, the effect of MTX may reduce this concurrent medication. Compared with other low oral dose studies, by weekly or cyclic dosing i.v. administration seems to show a better therapeutic effect on the synovial membrane, and at the same time better control of the patients is possible. Due to the i.v. administration and the documentation of the total dose, the possibility to adapt the therapy to the individual course of the disease of each patient is much better and the patient is able to show a higher degree of compliance. Moreover, possible side effects can be discovered at an early stage.

The MTX therapy is one of the most important basic therapies for PA. Due to inexpert use and control that often led to negative side effects this therapy seems to have fallen into discredit in rheumatology. With a low-dose therapy with intervals especially adapted to the individual course of the disease and a documentation of the total dose, similar to the gold therapy used in rheumatology, fewer side effects are to be expected with MTX medication.

References

1 Black, R.L.; O'Brien, W.M.; Scott, E.J. van; Auerbach, R.; Eisen, A.Z.; Bunium, J.J.: Methotrexate therapy in psoriatic arthritis. J. Am. med. Ass. *189:* 743–747 (1964).

2 Gubner, R.; August, S.; Ginsberg, V.: Therapeutic Supression of tissue reactivity. II. Effect of aminopterin in rheumatoid arthritis and psoriasis. Am. J. med. Sci. *221:* 176–182 (1951).

3 Klippel, J.H.; Decher, J.L.: Methotrexate in rheumatoid arthritis. New Engl. J. Med. *312:* 853–854 (1985).

4 Kragballe, K.; Zachariae, E.; Zachariae, H.: Methotrexate in psoriatic arthritis. A retrospectice study. Acta derm. vener., Stockh. *63:* 165–167 (1983).

5 Miller, J.A.; Dodds, H.J.; Lees, W.R.; Anderson, P.N.; Levene, G.M.; Munro, D.D.; Kirby, J.D.: A comparison of ultrasonography and liver biobsy in the assessment of methotrexate-induced hepatotoxicity in patients with psoriasis. Br. J. Derm., suppl. 24, p. 24 (1983).

6 Neild, V.S.; Joseph, A.E.; Marsden, R.A.; Fawcett, H.A.; Uff, J.S.: Liver ultrasound: A safe screening test for methotrexate patients. Br. J. Derm., suppl. 24, pp. 24–25 (1983).

7 Nyfors, A.; Poulsen, H.: Morphogenesis of fibrosis and cirrhosis in methotrexate-treated patients with psoriasis. Am. J. surg. Path. *1:* 235–243 (1977).

8 Nyfors, A.: Benefits and adverse drug experiences during long-term methotrexat-treatment of 248 psoriatics. Dan. med. Bull. *25:* 208–211 (1978).

9 Rees, B.: Methotrexate for psoriasis: 1981. Cutes *28:* 166–168 (1981).

10 Roberts, M.E.T.; Wright, V.; Hill, A.G.S.; Mehra, A.C.: Psoriatic arthritis. Ann. rheum. Dis. *35:* 206–212 (1976).

11 Roenigk, H.H., Jr.; Fowler-Bergfeld, W.; Curtis, G.H.: Methotrexate for psoriasis in weekly oral dosis. Archs. Derm. *99:* 86–93 (1969).

12 Roenigk, H.H., Jr.; Auerbach, R.; Maibach, H.I.; Weinstein, G.D.: Methotrexate guidelines. Revised. J. Am. Acad. Derm. *G:* 145–155 (1982).

13 Roy, G.G.; Auerbach, R.; Hillel, T.: Liver biopsies-liver scans in methotrexate-treated patients with psoriasis. Archs. Derm. *118:* 649–651 (1982).

14 Schilling, E.; Stadelmann, M.L.: Klinik und Röntgenmorphologie der Arthritis psoriasis. Colloquia Rheumatologica 18. Psoriasis arthritis, pp. 29–47 (Werkt. Verlag Dr. E. Banaschewski, München-Gräfelfing 1984).

15 Siegmeth, W.; Eberl, R.; Tausch, G.; Raith, W.: Vergleichende Untersuchung der Wirkung des Antimetabolismus Methatrexate bei Psoriasis, Arthropathia psoriatica und rheumatische Arthritis. Verh. dt. Ges. Rheumatol. *1:* 385–389 (1969).

16 Stadelmann, M.L.; Schilling, F.: Typeneinteilung der Arthritis psoriatica. Verh. dt. Ges. Rheumatol. *7:* 423 (1981).

17 Weinblatt, M.E.; Coblyn, J.S.; Fox, D.A.; Fraser, P.A.; Holdsworth, D.E.; Glass, D.N.; Trentham, C.B.; Trentham, D.: Efficacy of low-dose methotrexate in rheumatoid arthritis. New. Engl. J. Med. *312:* 818–822 (1985).

18 Wilkens, R.F.: Reappraisal of the use of methotrexate in rheumatoid arthritis. Am. J. Med. *October 31:* 19–25 (1983).

19 Wilkens, R.F.; Williams, H.J.; Ward, J.R.; Egger, M.J.; Reading, J.C.; Clements, P.J.; Cathcart, E.S.; Samuelson, C.O.; Solsky, M.A.; Kaplan, St.B.; Guttadauria, M.; Halla, J.T.; Weinstein, A.: Randomized, double-blind, placebo controlled trial of low-dose pulse methotrexate in psoriatic arthritis. Arthritis Rheum. *27:* 376–381 (1984).
20 Wright, V.; Moll, J.M.H.: Seronegative polyarthritis. (North-Holland, Amsterdam 1976).
21 Wright, V.: Psoriatic arthritis; in Kellaj et al. Textbook of rheumatology, chap. 66 (Saunders, Philadelphia 1981).

Dr. A. Dunky, Krankenhaus der Stadt Wien-Lainz, 2. Medizinische Abteilung mit Rheuma-Ambulatorium, Zentrum für Diagnostik und Therapie Rheumatischer Erkrankungen, Wolkersbergenstrasse 1, A–1130 Wien (Österreich)

Rheumatology, vol. 9, pp. 96–101 (Karger, Basel 1986)

Methotrexate Therapy of Reiter's Syndrome

Case Report

G. Herborn, R. Rau

Department of Rheumatology, Evangelisches Fachkrankenhaus, Ratingen, FRG

After the efficacy of methotrexate in psoriatic arthritis had been established, some reports about a few cases of Reiter's syndrome treated with methotrexate appeared, for instance by *Owen and Cohen* [2, 3] and *Luderschmidt and Balda* [1]. In a lecture about Reiter's syndrome in 1981, we presented 2 patients with this disease who had been treated successfully with methotrexate. Today, we wish to report about a patient with Reiter's syndrome who was treated with low-dose methotrexate and in whom this treatment was followed by a complete recovery from arthritis, urethritis and severe lesions of the skin and the nails.

In 1972, at the age of 30, the patient became acutely ill with all the symptoms of an acute Reiter's syndrome with severe urethritis, conjunctivitis, mucositis of the mouth and nose, and swelling of both knee joints and of the right ankle joint. In addition, there was a plantar exanthema with all characteristics of a keratodermia blennorrhagica. The patient was hospitalised in the Medical Department of the University Hospital of Düsseldorf, where he was treated with corticosteroids and indomethacin. After this treatment the patient recovered completely and was nearly free of symptoms for 6 years. During this time there were only slight arthralgias, from which he could be released by treatment with indomethacin.

Clinical Findings

In February 1978, the patient suffered from swelling of several joints and lesions on the skin and nails again. In February 1979, he came to our outpatient clinic for the first time: the clinical investigation revealed synovitis of all MCP joints of both hands with pain on pressure and movement, synovitis of the right wrist joint with limitation of movement, synovitis of the knee joints with effusion, synovitis of the left ankle joint and of all MTP joints.

The nails of the hands and feet showed a massive hyperkeratosis, subungual keratosis, onycholysis, redness and swelling of the end phalanges with a scaly skin and purulation (fig. 1a). There were disseminated scaling erythemas on both thighs. There was a severe balanitis (fig. 1b) as well as a rhinitis, whereas the conjunctiva were not affected.

The lumbar spine showed a painful restriction of movement, the same was true for the cervical spine. The paravertebral muscles were tense and painful. The mobility of the thoracical spine was 30/31.5 cm, of the lumbar spine 10/14 cm, finger-ground distance 20 cm (Schober's test), chest expansion 5 cm (difference of inspiratory and exspiratory).

Laboratory Findings

Sedimentation rate 90/125 mm, hemoglobin 12.8 g/dl, erythrocytes 4,200/mm^3, leukocytes 8,300/mm^3, platelet count 295,000/mm^3, total protein 65 g/l. Electrophoretic pattern: albumin 46%, α_1-globuline 5%, α_2-globuline 15%, β-globuline 14%, γ-globuline 10%. Rheumatoid factor and Rose-Waaler test negative, C-reactive protein positive, HLAB-27 positive. The other routine laboratory tests were normal with the exception of a slight elevation of alkaline phosphatase of 205 U/l (upper normal limit 170 U/l).

In the knee joint effusion 30,000 cells were counted, the fluid was sterile.

Radiological Findings

The radiological examination in February 1979 showed an osteoporosis adjacent to the joints of the wrist and the MCP joints. Some

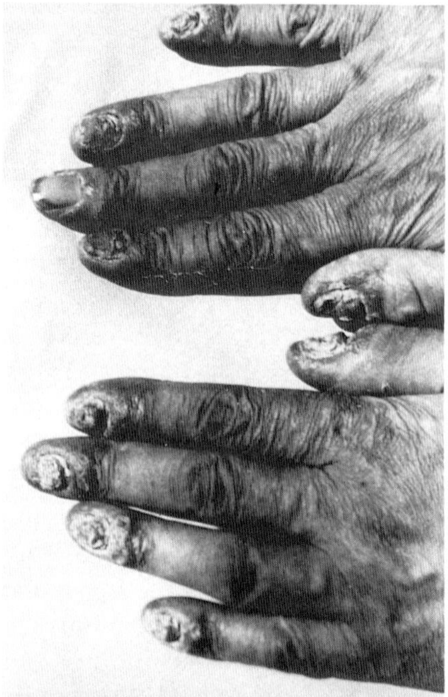

a

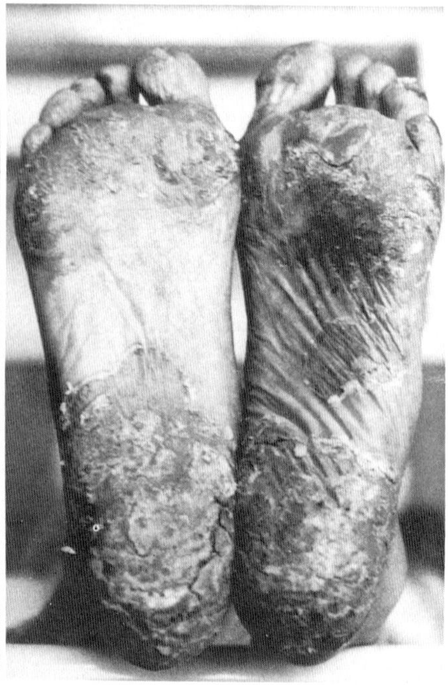

b

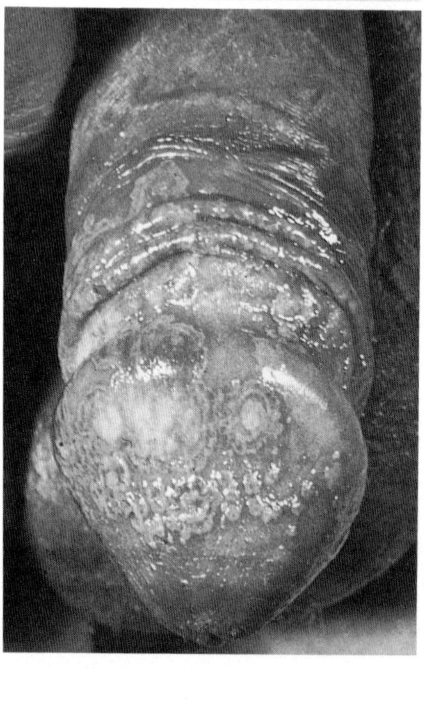

c

Fig. 1. The patient's nail changes *(a);* keratodermia blennorrhagica *(b)*; balanitis *(c)*, and sacroiliitis *(d)* before MTX was started. With MTX treatment skin and mucosal alterations cleared completely.

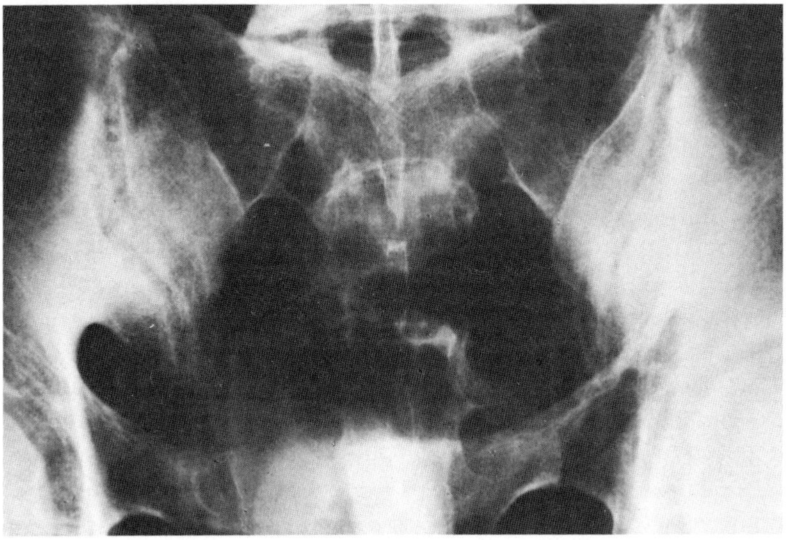

d

small cysts in the wrists could be recognized, but no erosions could be detected. The cervical spine was in a straight erect position. Between C6 and C7 beginning formation of atypical syndesmophytes. The X-ray of the lumbar spine showed squaring of the anterior vertebral margins (TH12 and 11) and reactive sclerosis. The sacroiliac joints showed a pseudowidening of the joint space, multiple erosions, irregular osteoporosis and sclerosis (fig. 1d).

Treatment and Course of the Disease

The patient was hospitalized in February, 1979. Because of the severity and chronicity of the disease, a remission-inducing treatment with gold salts was started, the patient received 150 mg of indomethacin per day and for a limited period of time 5 mg of prednisone every second day. Some intra-articular injections with triamcinolone acetonide were made and a radiosynovectomy of both knee joints with 5 μCi yttrium-90-silicate (5 μCi per knee joint) was performed.

Since with all these measures no definite improvement could be achieved, on April 6, 1979, treatment with methotrexate was started in addition to the gold treatment. The initial dose was 10 mg/week, later 25 mg/week. With this treatment, the pathological changes of the skin and nails and the joint symptoms cleared in a short period of time. The treatment was continued in our outpatient department until November 1979, when a total dose of 360 mg was reached.

During the following time the gold treatment was continued, the patient did well.

In April 1982, when the patient was seen again, an increasing limitation of spine movement attracted our attention. The X-ray control showed a new formation of atypical syndesmophytes C3/4 and C4/5, C6/7 were unchanged. Because of this rapid progression of syndesmophyte formation and because of a severe limitation of the mobility of the thoracical spine a radiotherapy of cervical and thoracical spine with 137-cesium with a total dose of 5 Gy was performed, physiotherapy and gymnastics were intensified. With this treatment a definite improvement of pain and mobility could be achieved.

One year later, in May 1983, the patient suffered from a new attack of his disease with severe changes of the skin, with keratodermia blennorrhagica (fig. 1b), nail destruction, balanitis, urethritis and painful swelling of both wrist joints, both elbow joints, both ankle joints and all MTP joints, in spite of the fact that the gold treatment had been continued. In addition to this treatment and in addition to 150 mg of indomethacin, methotrexate treatment was started again with 25 mg/week initially, and then 15 mg/week. Within 3 months the patient was again free of symptoms. As with former methotrexate treatment periods the synovitis responded faster than the skin and the nails.

We did not observe any reduction in efficacy of methotrexate treatment in this patient. After 5 months, in October 1983, a total dose of 255 mg was reached and the treatment could be stopped.

Since the patient continued to be free of symptoms and since a total dose of 4,500 mg of gold was reached, gold treatment was stopped in February 1984.

In April 1984, again the beginning of nail changes, slight balanitis and arthralgias of the right ankle joint and the right elbow joint were observed, and therefore the methotrexate treatment was resumed with 15 mg/week, some weeks later the dose could be reduced to 5 mg/week. Again, this treatment resulted in a total alleviation of symptoms.

References

1 Luderschmidt, C.; Balda, B.-R.: Morbus Reiter. Erfolgreiche Kombinationschemotherapie mit Methotrexat, Etretinat und Prednisolon. Münch. med. Wschr. *125:* 936 (1983).
2 Owen, E.; Cohen, M.: Methotrexate in Reiter's disease. Archs Derm. *115:* 716 (1979).
3 Owen, E.; Cohen, M.: Methotrexate in Reiter's disease. Ann. rheum. Dis. *38:* 48 (1979).

Dr. G. Herborn, Department of Rheumatology, Evangelisches Fachkrankenhaus, Rosenstrasse 2, D-4030 Ratingen (FRG)

Rheumatology, vol. 9, pp. 102–109 (Karger, Basel 1986)

Combined Chemotherapy of Reiter's Disease with Methotrexate, Aromatic Retinoid and Methylprednisolone

M. Knüchel[a], *B.R. Balda*[b], *Ch. Luderschmidt*[a]

[a]Universitäts-Hautklinik, Bonn-Venusberg; [b]Klinik für Dermatologie und Allergologie, Zentralklinikum, Augsburg, BRD

Introduction

In his original work in 1916, *Reiter* [22] reported about a soldier who, having recovered from diarrhea, developed a triad of symptoms, that is, urethritis, conjunctivitis and a painful swelling of the joints. In addition, we find in about 80% of cases hyperceratotic, erosive and psoriasiforme skin lesions [4, 11, 19]. Therefore, we nowadays speak of a tetrade in Reiter's disease [13]. Up to the present day the cause of the syndrome has not been established, but is considered to be a postinfection disease [2, 7, 16, 19].

Clinically, one must differentiate between the urethritic (or venereal) and the dysenteric types. The former variety begins with a so-called unspecific urethritis and is found predominantly in men. The ratio between men and women is 50 to 1.

The dysenteric type, as observed by *Reiter,* commonly starts with a Shigella dysentery. In this case a sex preference does not exist and it also occurs in children. About 2–4 weeks after an infection, urethritis, conjunctivitis, arthritis as well as psoriasis-like skin lesions develop.

At present, a satisfactory therapy of Reiter's disease does not exist. Apart from external preparations, aimed at giving symptomatic relief, glucocorticosteroids, gold preparations, antiphlogistic, immunosuppressive and cytostatic agents have been employed without convincing results [6, 9, 15, 21]. Currently, methotrexate has been widely reported to show the most favorable effect on Reiter's disease.

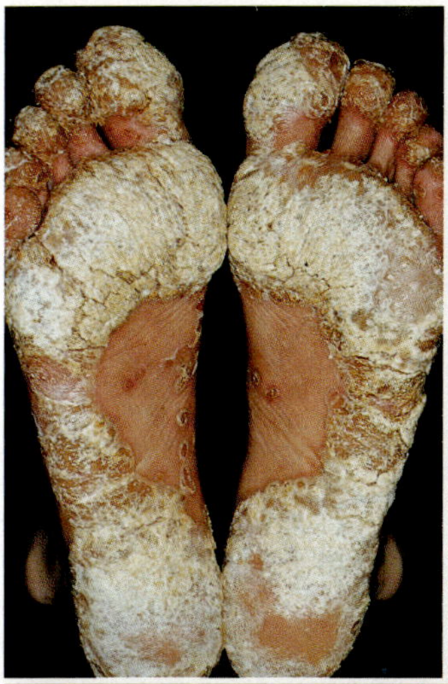
1

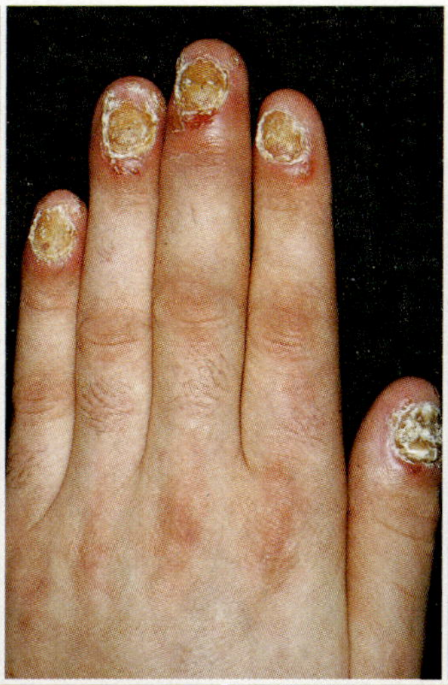
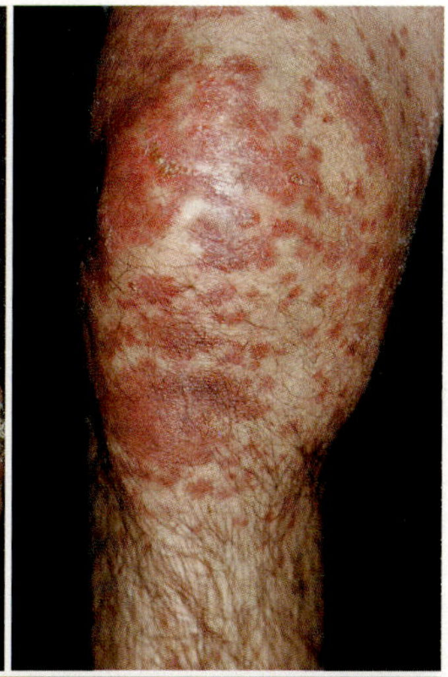
2, 3

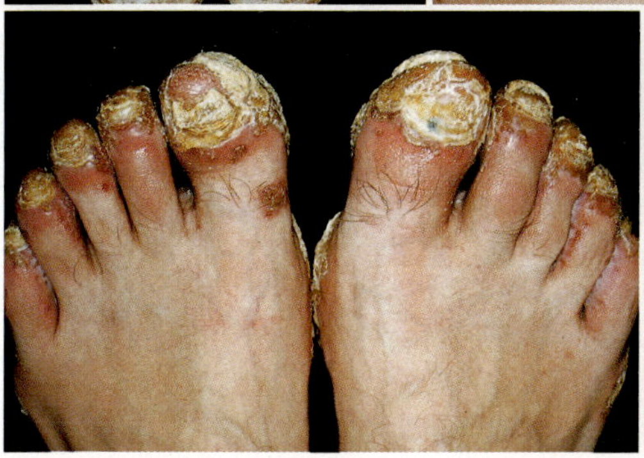
4

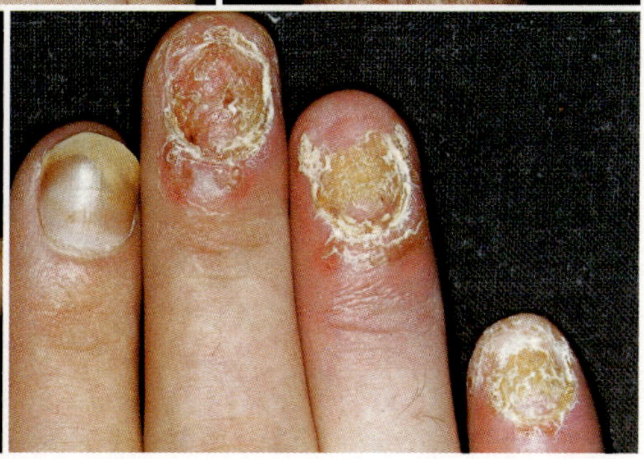
5

Fig. 1. Typical keratoderma of Reiter's disease; soles of patient 1.

Fig. 2. 'Krümelnägel'.

Fig. 3. Painful swelling of joints, typical psoriasiforme skin lesions.

Fig. 4. Inflammatory and exudative swelling of nail walls with typical nail changes.

Fig. 5. 'Krümelnägel' with subungual pustulation, pitting of nail plate and yellowish-brown oily spot in the nail of the second finger.

In the following report we would therefore like to present 3 men suffering from Reiter's disease aged 28, 30 and 36 years, whose condition was improved impressively by a combined therapy consisting of methotrexate, aromatic retinoid and methylprednisolone. In 1 case, the symptoms disappeared completely.

One patient started Reiter's disease with the unspecific urethritis, the 2 others with enteritis. After a lapse of 2–5 weeks they immediately developed severe asymmetric polyarthritis and the typical psoriasis-like skin lesions partially with pustules. Two patients were bedridden due to the severe involvement of the joints.

Case Reports

The first patient presented with typical discoid, shedding and pustulous lesions especially on the palms and soles (fig. 1). He showed erythematosquamous plaques extending from the face into the scalp as well as on the trunk and along the limbs, i.e. lots of nummular marginated livid-red centrally pitted succulent vesicular discs, which are found partly alone and partly confluent. Hyperkeratotic changes and as we say in German 'Krümelnägel' can be seen in the nails (fig. 2). Literally translated that means 'nails like crumbs'. This particular patient lost all his nails. Figure 3 shows the knee of this patient at the time of admission. The tissue above the joints is edematous and extremely painful.

The second patient, 30 years of age, suffered from an unspecific urethritis with a white discharge and painful swelling and reddening of the left foot. Four weeks later he developed pustules and a keratosis blennorrhagica on his palms and soles. The capillitium showed large erythematosquamous plaques. Succulent discoid lesions with scaling crusts could be found loosely distributed on the trunk and the limbs. All nail plates had been destroyed and again looked like 'Krümelnägel'. This particular patient did not show any signs of conjunctivitis, but complained of severe pain affecting the larger joints.

The last patient under review was admitted with the first recurrence of Reiter's disease. He presented with the characteristic symptoms of enteritis, transitory conjunctivitis, keratosis blennorrhagica, and painful involvement of the joints of his right knee, his left elbow as well as of his shoulders some time later on. The skin lesions of this patient have mostly been on the limbs. The palms and soles are covered by a thickened red and scaly keratotic layer, a keratoderma, which extends over the edges on to the dorsal surface of the fingers and toes. The endphalanges of fingers and toes are swollen with inflammatory infiltrated nail walls (fig. 4). The nail plates are split, shedding and crushed, partly with subungual pustulation on the right second finger (fig. 5).

In all 3 cases, histopathology shows a massive ortho-para-hyperkeratosis with embedded neutrophilic abscesses. Polymorphonuclear leukocytes lie in the rete Malpighi (fig. 6). We see elongation of the rete ridges, edematization and cogestion of blood at the top of the papillaries, and, in the upper corium slight lymphohistiozytic infiltration. In

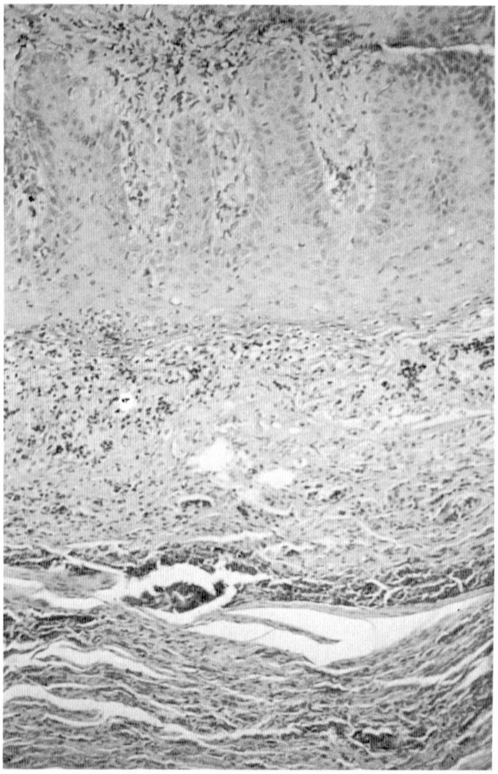

6

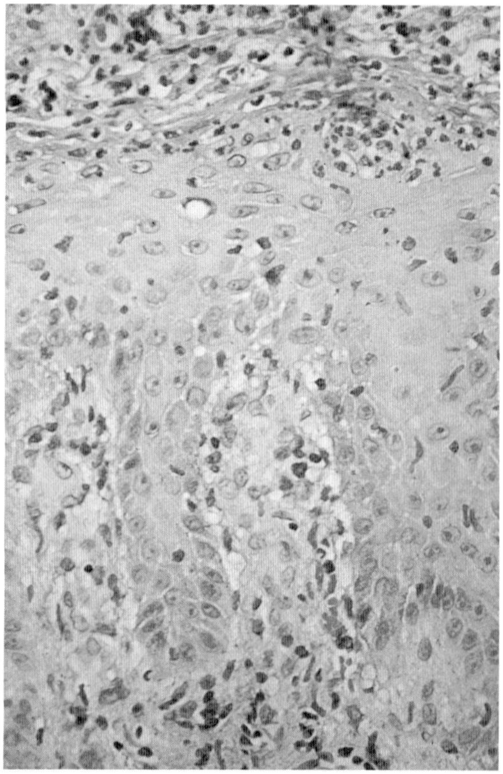

7

Fig. 6. Histopathology of Reiters' disease: ortho-para-hyperkeratosis with embedded neutrophilic abscesses. Polymorphonuclear leukocytes in the rete Malpighi. Elongation of the rete ridges, edematization and congestion of blood at the top of the papillaries. In the upper corium, slight lympho-histiocytic infiltration. HE. × 40.

Fig. 7. Histopathology in detail: Lots of polymorphonuclear leukocytes in the stratum corneum, elongated rete ridges, congestion of blood in the vessels and inflammatory round cell infiltrates. HE. × 100.

detail (fig. 7): Lots of polymorphonuclear leukocytes and a developing Munro-abscess in the stratum corneum; elongated rete ridges; congestion of blood in the vessels and inflammatory round cell infiltrates. In the first case, direct immunofluorescence demonstrated complement C_3 perivascularly and in the area of the basal membrane.

At present, no laboratory parameters exist for the diagnosis of Reiter's disease. In all patients an increased erythrocyte sedimentation rate and a polymorphonuclear leukocytosis could be demonstrated (table I). In 2 cases, serum electrophoresis showed a shifting. HLA-B27 [1, 14] could be demonstrated in all patients (table II). The rheuma-

Table I. Laboratory findings: inflammatory signs

Patient No.	ESR	Leukocytosis	$Alpha_2$-globulin
1	120/150	18,900	17.5 rel. %
2	85/111	15,600	normal
3	77/117	16,500	15.5 rel. %

Table II. Laboratory findings

Patient No.	HLA-B 27	AST	ANA
1	positive	600 Iu ↑	negative
2	positive	negative	negative
3	positive	negative	negative

Table III. Combined chemotherapy

Treatment
Methotrexate 50 mg/week i.v.
Methylprednisolone 60 → 12 mg/die orally
Aromatic retinoid (Tigason®) 75 → 25 mg/die orally

toid factor is usually not present and was positive in only 1 patient. Antinuclear antibodies and circulating immune complexes were negative in all 3 cases.

Neither the bacteriological nor the virological test of the serum, the urethral discharge, nor the aspiration fluid of the knee were of any diagnostic value.

In sharp contrast to the excrutiating pain experienced by the patients, X-ray investigation of the affected bones and joints completely failed to establish a cause.

Treatment (table III)

All our patients received the same treatment. Initially, they received a daily dose of 75 mg of aromatic retinoid (Tigason®) and 60 mg methylprednisolone orally and once per week 50 mg of methotrexate intravenously. In the third week, the dose of retinoid was reduced to 50

8 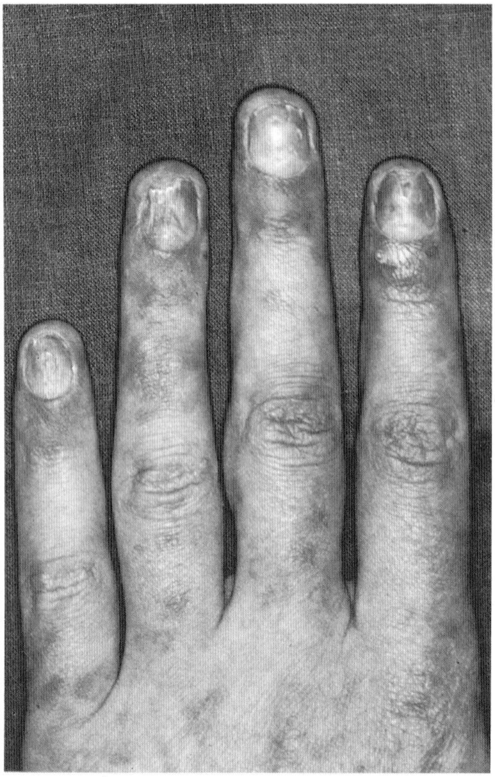

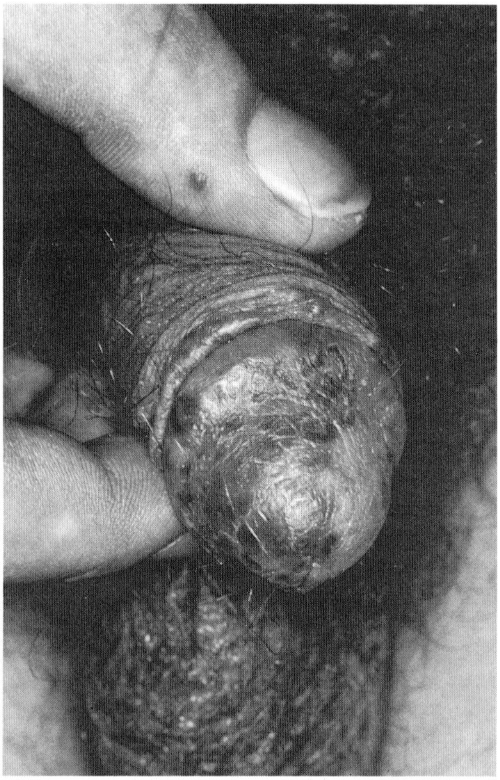 9

Fig. 8. Nails, 3 months after the start of therapy.

Fig. 9. Penile erosions, 5 weeks after the start of therapy.

mg/day. In accordance with the subsiding signs of inflammation and joint involvement, the dosage of methylprednisolone was gradually reduced to 12 mg on alternate days in the fourth and fifth weeks of therapy to suit the needs of the individual patient. It was discontinued in the eighth week of therapy.

Under this treatment, the condition of all the patients improved dramatically and after 6–10 weeks they could be discharged nearly free of symptoms. In 2 patients, we initially observed a transient rise of transaminases; after 4 weeks, however, we found a return to normal. Here we must remember that aromatic retinoid and methotrexate are both toxic agents to the liver [5, 8, 10, 12, 17, 23]. On leaving the hospital, all patients were also nearly free of skin symptoms. The nails grew healthy (fig. 8). Mucosal lesions were healing (fig. 9). As outpatients, however, the men continued treatment with 25 mg methotrexate per week and 50–25 mg aromatic retinoid per day.

In all the men, only the combined therapy was effective. Prior to admission to our hospital the first patient had been given 80 mg of methylprednisolone in combination with methotrexate at a dose of 15 mg daily without any success. Another received penicillin and indomethacin without any effect.

Discussion

Up to the present day the cause of the syndrome has not yet been established. When the treatment is based on symptoms, the psoriasis-like skin lesions and the keratosis blennorrhagica require different therapy than the joint symptoms with humoral signs of inflammation [20, 26].

Especially etretinate and methotrexate are important drugs for treating the pustulous kind of psoriasis vulgaris (type v. Zumbusch) and psoriasis arthropatica. Aromatic retinoid has a dose-related antikeratotic effect by diminishing new keratin formation and the contact of keratinocytes [24] as well as a modulating effect on immune competent cells. Methotrexate, however, not only reduces an increased cell proliferation [4] in psoriatic epidermis [25], but also prevents the dose-related chemotactic migration of polymorphonuclear leukocytes, which plays an important role in the psoriasis pustulosa and pustulous eruptions of Reiter's disease [3, 18].

Our therapeutic approach was intended to influence the existing clinical and laboratory findings symptomatically. We came to the conclusion that the extent of side effects was reduced to a bearable level by this combination. So our expectations were fully met. We even noticed a hyperadditive effect of the single components of this triple combination so that a relatively short period resulted in a satisfactory success of therapy.

Finally, we want to emphasize that this treatment is not general therapeutic advice but must be reserved for severe cases. Therefore, more experience must be searched for.

References

1 Calin, A.: An 'experimental' epidermic of Reiter's syndrome revisited. Follow-up evidence on genetic and environmental factors. Ann. intern. Med. *84:* 564–566 (1976).

2 Catterall, R.D.: The role of microbial infection in Reiter's syndrome; in Dumond, Infection and immunology in the rheumatic diseases, pp. 147–150 (Blackwell, Oxford 1976).

3 Farber, G.A.; Forshner, J.G.; O'Quinn, S.E.: Reiter's syndrome. Treatment with methotrexate. J. Am. med. Ass. *200:* 181–183 (1967).

4 Gelfaut, S.: The cell cycle in psoriasis. A reappraisal. Br. J. Derm. *95:* 577–590 (1976).

5 Glazer, S.; Sparberg, M.; Yakoo, H.; Roenigk, H.H., Jr.: Ro 10-9359 in psoriasis Study of effectiveness and potential hepatotoxicity (Abstract). J. invest. Derm. *76:* 303 (1981).

6 Good, A.E.: Reiter's disease: modern concept. Cutis *24:* 514–523 (1979).

7 Good, A.E.; Schultz, J.S.: Reiter's syndrome following Shigella flexneri. Arthritis Rheum. *10:* 100–105 (1977).

8 Grupper, C.; Berretti, B.; Bermejo, D.; Gougne, B.: Maladie de Reiter. Considérations évolutives et thérapeutiques (Méthotrexate, PUVA-thérapie et rétinoïde aromatique): 3 cas, Journées dermatologiques, Paris 1980.

9 Harris, J.W.R.: Recent advances in sexually transmitted diseases, in Morton, Harris, p. 314 (Churchill-Livingstone, Edinburgh 1975).

10 Lassus, A.: Systemic treatment of psoriasis with an oral retinoid acid derivation (Ro-10-9359). Br. J. Derm. *1980:* 102–195.

11 Luderschmidt, Ch.; Balda, B.-R.: 'Morbus Reiter'. Erfolgreiche Kombinations-Chemotherapie mit Methotrexat, Etretinat und Prednisolon. Münch. Med. Wschr. *125:* 936–940 (1983).

12 Mahrle, G.; Orfanos, C.E.; Ippen, H.; Hofbauer, M.: Haarwachstum, Leberwerte und Lichtempfindlichkeit unter oraler Retinoid-Therapie bei Psoriasis. Dt. med. Wschr. *104:* 473–477 (1979).

13 Montgomery, M.M.: The microcutaneous lesions of Reiter's syndrome. Ann. Intern. Med. *51:* 99–109 (1959).

14 Morris, R.; Metzger, A.L.; Bluestone, R.; Terasaki, P.J.: HLA-B 27. A clue to the diagnosis and pathogenesis of Reiter's syndrome. New Engl. J. Med. *290:* 554–556 (1974).

15 Mullins, J.F.; Maberry, J.D.; Stone, O.J.: Reiter's syndrome treated with folic acid antogonists. Archs Derm. *94:* 335–339 (1966).

16 Noer, H.R.: An 'experimental' epidemic of Reiter's syndrome. J. Am. Med. Ass. *198:* 693–698 (1966).

17 Orfanos, C.E.; Mahrle, G.; Goerz, G.; Happle, R.; Hofbauer, M.; Landes, E.; Schimpf, A.: Laboratory investigations in patients with generalized psoriasis under oral retinoid treatment, Dermatologica, Basel *159:* 62–70 (1979).

18 Owen, E.; Cohen, M.: Methotrexate in Reiter's disease. Archs Derm. *115:* 716–718 (1979).

19 Paronen, I.: Reiter's disease: A study of 344 cases observed in Finland. Acta med. scand. *212:* suppl., pp. 1–114 (1948).

20 Perry, H.O.; Mayne, J.G.: Psoriasis and Reiter's syndrome. Archs Derm. *92:* 129–136 (1965).
21 Perry, H.O.: Reiter's syndrome; in: Fitzpatrick, Dermatology in general medicine, pp. 1327–1331 (McGraw Hill, New York 1979).
22 Reiter, H.: Über eine bisher unbekannte Spirochäteninfektion (*Spiroaetosis arthritica*). Dt. med. Wschr. *42:* 1535–1536 (1916).
23 Robinson, J.K.; Baughman, R.D.; Auerbach, R.; Cimis, R.J.: Methotrexate hepatotoxicity in psoriasis. Archs Derm. *116:* 413–415 (1980).
24 Schultz-Ehrenberg, A.; Orfanos, C.E.: Light and electron-microscopic changes of human epidermis under oral retinoid treatment; in Orfanos, Retinoids, 85–92 (Springer, New York 1981).
25 Weinstein, G.D.: Methotrexate. Ann. intern. Med. *86:* 199–204 (1977).
26 Wright, V.; Reed, W.B.: The link between Reiter's syndrome and psoriatic arhritis. Ann. rheum. Dis. *23:* 12–20 (1965).

Dr. Martin Knüchel, Universitäts-Hautklinik, Sigmund-Freud-Strasse 25, D-5300 Bonn 1 (BRD)

Rheumatology, vol. 9, pp. 110–112 (Karger, Basel 1986)

Discussion 4

Karger: A question to Dr. *Nyfors:* apart from written recommendations, do you perform needle biopsy of liver in every patient before starting MTX therapy?

Nyfors: Yes.

Rau: I wish to ask Dr. *Nyfors* about the average weekly doses he used in his dermatological patients. These doses seemed to me to be relatively large. In our patiens we got the impression that at least some side effects are dose-related. Did you find any correlation between dose and side effects?

Nyfors: The maintenance dose was around 10–15 mg, and it is obvious that a parameter like nausea, is dose-related. So, when we gave 25 mg in the beginning, they had nausea. When we came down and passed 15 mg per week on the way down, the nausea often disappeared. So the mean dose was 10–15 mg.

Furst: Dr. *Nyfors,* you mentioned that salicylates were a contraindication to the use of methotrexate. Relative, absolute? How much salicylate? Any salicylate?

Nyfors: No, I didn't say that it was a contraindication. I think I had a slide showing what we ask at pre-MTX evaluation. We allowed them to have salicylates, but we usually ask them not to use them the 2 days they have methotrexate 'in the body', as we express that. So they could have it the remaining 5 days of the week.

Jürgens: I have some questions for Dr. *Nyfors* relating to the length of treatment. How long did you treat your patients and when you stopped treatment did you see any sustained remissions? And the other question is: did you see any relapses on treatment indicating there some sort of resistance to methotrexate was developing? And then the next question is, the mechanism of action, did I understand you right, as a pediatrician I might be allowed the question that in psoriasis there is an increased turnover in the lower skin layers of cells, so that the mechanism of action could be related in this disease to the antiproliferative action of methotrexate as in contradiction to the mechanism of action in rheumatoid arthritis. And if so, is it a different mechanism of action in skin lesions as compared to the arthritis?

Nyfors: That is certainly a lot to answer. We treat the patients as long as they need the methotrexate. Usually, psoriasis has a mild course during the summer, so they have a methotrexate-free period during summer, and we have been up to 12,000 mg in some patients. I will come back tomorrow on how to follow them with liver biopsies and that sort of thing. We discontinued methotrexate in some patients because we saw relapses. The first time I made a study of the results of methotrexate therapy in psoriasis vulgaris in my patients was in 1970 and then I made a similar evaluation in 1978, and the results of complete clearing were 62% in the first study and 52% in the second. So some patients seem to have experienced less effect with methotrexate during this long-term therapy. And the third question, regarding the antiproliferate effect of methotrexate. We certainly believe in what we dermatologists use for our basal layer. I don't think we have any, or use the immunosuppressive effects of methotrexate. Is that some sort of an answer?

Miescher: With a psoriatic arthritis, the question is, are we dealing with two different mechanisms or is it the same mechanism? I think we do not yet have a solution today. It might be of interest to you that I had a patient with psoriatic arthritis, the psoriasis was not terribly important, but the arthritic complaints were sufficient to warrant treatment. And this patient received levamisole which cleared the arthritis without any influence on psoriasis. So at least, with regard to levamisole, we could say we had an influence on the joint problem without any influence on the skin problem.

Weinstein: I have a question for Dr. *Herborn.* You used both gold and methotrexate, you mentioned. I think there is the impression also that in psoriatic arthritis gold can be an effective agent but does nothing for the rash.

Herborn: Yes, gold is an effective agent in the treatment of the psoriatic arthritis but does not improve the psoriatic rash.

Weinstein: Have you used gold much in Reiter's syndrome? Does it also help the arthritis and not do much for the skin rash?

Herborn: Yes, we have used gold in Reiter's syndrome. It has a good effect on the arthritis and does not change the skin rash.

Weinstein: Do you think that methotrexate alone without gold might help both the skin and the joints, so you wouldn't have to use gold at all? I didn't quite understand why you were using both – gold didn't appear to be at all effective here.

Herborn: We treated with both drugs because of the severity and chronicity of the disease. Because gold salts didn't show sufficient efficacy.

Weinstein: Yes, but the question was, if gold wasn't doing anything here. I had the impression that the gold wasn't doing anything.

Rau: I think in this case we started with the gold for the arthritis. But, on the other hand, there were very severe skin and nail lesions, and so we tried to treat this aspect of the disease as well, and for this we used methotrexate. We have used this combination also in patients with rheumatoid arthritis, for our philosophy to use gold as as we say in German remission-inducing treatment for long time, and use methotrexate only for short periods during which the patient had a flare or a very high activity of his disease. So, I think, the gold is basically influencing the disease, and methotrexate is used in addition to gold, to give some sort of symptomatic relief. I think that was the idea. If the patients tolerated gold we did not see more or more serious side effects when combining MTX with gold than with MTX alone, but since we believe that both drugs have a different way of action we hope to get a better or more complex effect on the disease than with the single drug.

Wilke: I have one question for Dr. *Nyfors.* I understood that you give your methotrexate in one dose. *Weinstein,* a few years back suggested that the best way to give methotrexate to control psoriasis because of epidermal cell half-life was a 12-hour times 3. And I know that all my colleagues continue to give methotrexate that way in treating rheumatoid arthritis. Why did you choose to give methotrexate in one dose?

Nyfors: It has been shown by *Roenigk* in a multicenter study, that the hepatotoxicity of these 3 divided doses is larger than if you give a single dose. So, from the beginning we refused to use the *Weinstein* scheme, because we have to prevent our patients from the side effects in the long run. They are more important than the effect. And we have done a single weekly study so long that we knew that this was effective with one single dose. So why should we change over to the divided dose? And we haven't.

Williams: I have a question for Dr. *Herborn:* You apparently used gold in combination with methotrexate in other problems. Did you find an increase in toxicity by using the two of them together?

Herborn: We did not yet evaluate the side effects of patients who had the combination of gold with MTX and who had MTX without gold statistically. But our impression is that there is no increase of toxicity when combining both drugs.

Rheumatology, vol. 9, pp. 113–115 (Karger, Basel 1986)

Methotrexate Therapy in Patients with Rheumatoid Arthritis

D. Gross

Klinik für Rheumatologie und Rehabilitation, Stadtspital Triemli, Zürich, Schweiz

Three substances were on trial 20 years ago at the University Hospital of Rheumatic Diseases in Zurich to assess their effectiveness in patients with rheumatoid arthritis. We used systemically: mercaptopurin (purinethol) as a purin antagonist; azathioprin (imuran) as a purin antagonist, and amethopterin (methotrexate) as a folic acid antagonist.

The goal of the therapy was to influence the immunological system to block a part of its reactions and to prove the efficiency of the drug as a new basic treatment for rheumatoid arthritis. 35 patients (25 women, 10 men) were treated by intravenous injection of 0.05 mg methotrexate once a week.

31 patients had classical rheumatoid arthritis stages II–IV, 2 patient had polysynovitis in lupus erythematodes disseminatus and 2 patients had periarteriitic symptoms in addition to the polysynovitis. This means that the patients were severely ill and hard to treat. 31 patients were under corticosteroid treatment for at least 6 months.

Methods

Methotrexate was given at a dosage of 50 mg i.v. once a week for a minimum of 6 weeks and a maximum of 20 weeks. The dosage was influenced by and adapted to that in the publication of Dr. *Miescher*, New York, who treated patients with lupus erythematodes disseminatus in the same way.

20 patients have been treated with methotrexate alone. 9 patients received after the first 6 weeks of methotrexate alone, a combination of methotrexate intravenously and azathioprin orally, 5 patients received a combination of methotrexate and 6-mercaptopurin, and 1 patient received a combination of all three substances together. The results were published in 1967 [1, 2], and I can give you a short report of them.

From the clinical point of view, 22 patients had marked improvement of their general condition within the first 6 weeks, when methotrexate has been given alone, 9 patients did not show any change in their general condition and 4 patients showed an aggravation of their condition.

This was why after 6 weeks of monotherapy treatment with a combination of methotrexate and purinethol or imuran was attempted.

Results

The sedimentation rate dropped in 16 patients after the 6-week period, was unchanged in 7 patients, and had a tendency to increase in 12 patients. However, in all these patients the local inflammatory symptoms of synovitis in the affected joints did not change markedly.

Special attention has been paid to the dosage of corticosteroids during the methotrexate trial. 31 patients had taken at the beginning of the antimetabolic therapy, corticosteroids for at least 6 months. In 22 patients the dosage could be reduced within the first 6 weeks and, later, in 5 patients the corticosteroids could not be changed and in 4 patients the corticosteroids had to be raised to a dosage of 200% of the initial one.

It is noteworthy that the 4 patients without corticosteroids at the beginning of the trial had improvement of their general condition, drop in sedimentation rate, and functional improvement.

The rheumatoid factor in serum was positive in 31 patients with titers in the Singer and Plotz test of between 1:80 and 1:2,560. 4 patients became seronegative during the trial, 9 patients had a drop of in titer of more than 2 degrees, 20 patients reminded unchanged and 2 patients had aggravations.

In 7 patients with antinuclear antibodies (antiglobulin consumption-test) the test became negative.

No significant changes in the complement reactions of Ch50, C3 and C5 could be found.

Side Effects

The peripheral leukocytes dropped in 25 patients by an average of 3,336 and increased in 10 patients by an average of 2,214. The maximal drop of leukocytes was 13,500 in a patient who died at agranulocytosis and gastrointestinal infection.

In addition, the leukocytes count dropped in 4 patients in the peripheral blood to under 3,000, so that methotrexate treatment had to be stopped.

As additional side effects, we noted 2 allergic skin reactions, 4 instances of stomatitis, 6 of nausea and, at the end of the combined trail, 4 of gastrointestinal incompatibilities.

Conclusions

In conclusion our impressions about methotrexate in 35 patients with rheumatoid arthritis and allied conditions were: (1) Methotrexate has some efficiency in rheumatoid arthritis concerning the general condition of the patients, improvement of sedimentation rates, and the possibility of reducing the dosage of corticosteroids. (2) Methotrexate given as an intravenous injection of 50 mg once a week is too dangerous due to the severe side effects in the bone marrow (agranulocytosis).

The influence of serological reactions like rheumatoid factors, antinuclear antibodies, and complement fractions are difficult to explain because of their high variability with no uniform reactions.

References

1 Gross, D.; Enderlin, M.; Fehr, K.: Die immunsuppressive Behandlung der progredient chronischen Polyarthritis mit Antimetabolica und Cytostatica. Schweiz. med. Woschr. *97:* 1301–1310 (1967).

2 Gross, D.; Enderlin, M.: Beitrag zur Behandlung der progredient chronischen Polyarthiritis mit Antimetaboliten und Cytostatica. Z. Rheumaforsch. *26* (1967).

Prof. Dr. D. Gross, Klinik für Rheumatologie und Rehabilitation,
Stadtspital Triemli, Birmensdorferstrasse, CH-8063 Zürich (Schweiz)

Rheumatology, vol. 9, pp. 116–122 (Karger, Basel 1986)

Experience in the Use of Intravenous Methotrexate Therapy in Rheumatoid Arthritis

Robert M. Michaels

Chief-Department of Rheumatology, Guthrie Clinic Ltd., Sayre, Pa., USA

I first became interested in using methotrexate in patients with rheumatoid arthritis during my rheumatology fellowship at Georgetown University in about 1978. At that time I had been following several patients with polymyositis whose disease had been refractory to high dose steroid therapy and we were treating them with weekly intravenous methotrexate at an average dose of 50 mg. This of course was and still is considered standard effective treatment for steroid-resistant polymyositis, and I was impressed with its effectiveness in this disease as well as how well the patients seemed to tolerate it. Of course we do not know the mechanism of how methotrexate works in polymyositis, but as naive as I was then and perhaps still am now, I postulated that if methotrexate were working in these patients by killing the lymphocytes attacking their muscles, why wouldn't it also kill the lymphocytes attacking the synovium in patients with rheumatoid arthritis. I also had a number of patients in my clinic with active rheumatoid arthritis who had not responded to hydroxychloroquine and gold salts and who were candidates for therapy with either *D*-penicillamine or immunosuppressive agents, the most common ones in use at that time being azathioprine, cyclophosphamide, and chlorambucil. Acute and chronic toxicity with all of these agents was and still is considerable and caused us to have second thoughts about using these agents. In addition, a number of patients refused to go on these immunosuppressive agents because of their carcinogenic potential. Studies in patients treated with chronic methotrexate therapy, however, have not shown an increased incidence of malignancy, and it was believed back in 1977 and still is now that methotrexate is not carcinogenic [1, 2].

In view of this and the favorable experience with methotrexate in polymyositis, I wondered why it had not been used in rheumatoid arthritis. I then went to the literature to see if anyone had tried it in rheumatoid arthritis and was able to find a couple of studies. The first one was done by *Gubner* et al. [3] and published in 1951 in the American Journal of Medical Science. They studied 6 patients with rheumatoid arthritis, 1 with psoriatic arthritis and 1 with acute rheumatic fever. In this study, oral aminopterin was given in a dose of 1–2 mg/day for 6–21 days. Five out of 6 of their rheumatoid arthritic patients improved on this therapy and did so rather quickly with the first signs of improvement being noted between 3 and 17 days and maximum improvement occurring in 7–23 days. How much improvement these patients obtained is not known since there was no quantitative information given in the paper.

The second study I was able to find was in the German literature by *Gross* et al. [4]. They studied 35 patients with connective tissue diseases – 31 with rheumatoid arthritis, 2 with polyarteritis nodosa, and 2 with systemic lupus. 50 mg of methotrexate was given intravenously for 6 consecutive weeks and the patients were then switched and maintained on azathioprine or another purine antagonist. Of the 35 patients, 22 improved and the improvement seemed to occur over several weeks.

Because so little had been written about the use of methotrexate in rheumatoid arthritis and because methotrexate was potentially safer than many of the other immunosuppressive agents previously used to treat rheumatoid arthritis, and, finally, because of some evidence from these early studies that methotrexate might very well be effective in treating rheumatoid arthritis, we decided to do a study with the major aim being to determine whether the methotrexate would be effective in inducing suppression of synovitis in patients with rheumatoid arthritis who had been refractory to standard therapy with at least hydroxychloroquine and gold salts.

We chose the intravenous relatively high dose method of administration for several reasons. First, we were comfortable in using it this way since we had been giving it by this method to patients with poly myositis and had not incurred any serious toxicity. Second, the main purpose of the study again was simply to determine if methotrexate would be effective in refractory rheumatoid arthritis, and we wanted to give enough of it so that if it did not work we would not wonder whether it was simply because we did not give enough. We really did

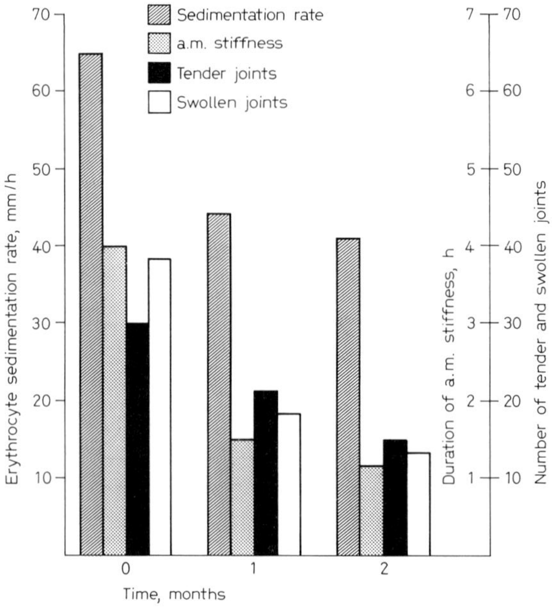

Fig. 1. Change in parameters of disease activity during methotrexate therapy.

not want to go much higher than 50 mg because in nonmalignant diseases, doses higher than this had not been used extensively and we did not want to get into problems with severe toxicity. We also felt that the intravenous route would give us total control of the drug and compliance would not be a problem. And, finally, we envisioned that if in fact the drug worked, we could perhaps gradually reduce the frequency of administration to every 2 weeks for a couple of months then every 3 weeks for a couple of months and then maintain the patient with one intravenous shot every 3 or 4 weeks in much the same way that gold injections are administered.

Although I wrote the protocol for the study during my fellowship in 1978, the project really did not get started until early 1980 after I had started practicing clinical rheumatology at the Guthrie Medical Center. It was a cooperative effort and of the 14 patients that were finally entered, 6 were my own, 6 were contributed by Dr. *David Nashel* at the Washington D.C. Veterans Administration Hospital, and 2 were entered by Dr. *Steve Derbes*, a rheumatologist in New Orleans, La., and a former Georgetown fellow.

All of the 14 patients in this study met the American Rheumatism criteria for classic rheumatoid arthritis; all of them had clinical evidence of active disease; and all of them had failed to respond to both gold salts and hydroxychloroquine. In addition, 6 patients were unresponsive to *D*-penicillamine, 1 to azathioprine and 1 to cyclophosphamide. All the patients were allowed to continue their nonsteroidal anti-inflammatory drugs and/or corticosteroids in a stable dosage throughout the study.

Methotrexate was administered intravenously at weekly intervals. We initially gave a 'test dose' of 10 mg followed by one 25-mg dose and then 50 mg intravenously weekly. Ten of the 14 patients were able to tolerate this schedule for at least 2 months, and 4 required dose modification or withholding of the drug either because of oral ulcers, liver enzyme abnormalities, or nausea and vomiting. Nausea actually was a common side effect being seen in 8 of 14 patients usually within 24 h of injection, but only once was it severe enough to cause discontinuation of treatment during the study. No patient experienced leukopenia, thrombocytopenia, increased anemia, or changes in blood urea nitrogen or creatinine.

Eleven out of the 14 patients in the study showed significant subjective and objective improvement within 2 months of starting methotrexate therapy and most did so by 1 month. Figure 1 illustrates the changes that occurred in the Westegren sedimentation rate, duration of morning stiffness, and joint index measured by the number of tender and swollen joints in the patients who responded to methotrexate. There was significant improvement in all of these parameters during the 2 months that the patients received the drug.

So the answer to the question of whether or not methotrexate would effectively treat refractory rheumatoid arthritis seemed to be 'yes'. Although this was not a double-blind study, we felt that it would have been very unlikely that improvement would have occurred spontaneously in such a large percentage (over 78%) of this group of patients with hitherto very resistant disease. These are the results that we published in *Arthritis and Rheumatism* in March 1982, and our formal study ended here [5].

The problem we now had was what to do with the patients who responded favorably. As I mentioned earlier we had envisioned the ideal situation to be one where intravenous methotrexate could be administered in much the same way as parenteral gold by decreasing the frequency of injections to a maintenance schedule of every 3–4 weeks.

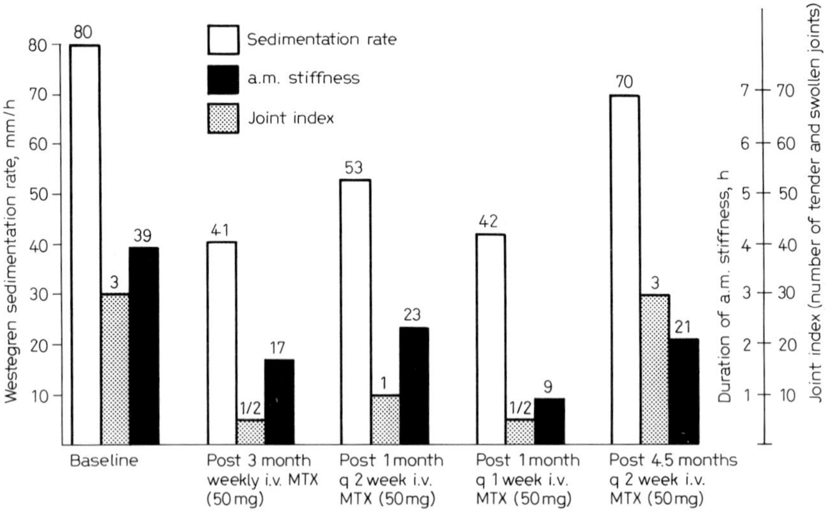

Fig. 2. Patient number 4. Response to methotrexate (MTX) tapering.

Unfortunately, however, what both Dr. *Nashel* and I found is well illustrated by the course of this particular patient who was typical for the group in general (fig. 2). As the frequency of the methotrexate injections was reduced even to every 2 weeks, the disease tended to flare over a variable period of time. It also seemed to respond well and rapidly to reinstitution of the weekly intravenous schedule. So, in this patient for example, a significant improvement was noted after receiving weekly intravenous injections for 3 months, but after only 1 month of every 2-week injections, her disease started to flare and then was again promptly brought under control after 1 month of weekly intravenous injections. Her disease then flared again almost to baseline levels after 4.5 months of every 2-week injections.

By this time, Dr. *Willkens* from Seattle and Dr. *Wilke* from Cleveland Clinic had published their favorable experience with low-dose pulse oral methotrexate in patients with rheumatoid arthritis [6,7]. The majority of our patients were then gradually switched from the relatively-high-dose weekly intravenous route to the low-dose weekly oral route. Of the original 11 responders, 6 have maintained a prolonged and good remission on 5–20 mg of oral methotrexate weekly. One patient did well on 7.5 mg of oral methotrexate for about 2 years, but his disease flared apparently spontaneously and not due to lack of compliance. He was subsequently started back on 25–35 mg of weekly intravenous methotrexate and his arthritis and general condition has improved. He continues on weekly intravenous methotrexate and is doing well.

Four patients were changed from intravenous methotrexate to other medications because of toxicity. Two are on azathioprine and doing well. One was tried on chlorambucil, then *D*-penicillamine, and then azathioprine, none of which helped his arthritis. He is reconsidering going back on weekly intravenous methotrexate. The last patient had been on gold, hydroxychloroquine and *D*-penicillamine prior to entering the methotrexate study. He discontinued the methotrexate because of nausea and vomiting after the study was completed. He took one dose of azathioprine, developed nausea and vomiting and refused to take any more. Total lymph node irradiation was tried but again he developed nausea and vomiting after 1,250 rad were delivered over 2 weeks. This was discontinued and he was maintained on 20 mg of prednisone a day alone and was lost to follow-up.

In summary, we can conclude that weekly relatively-high-dose methotrexate is effective in treating patients with refractory rheumatoid arthritis. Whether it is more effective than lower oral doses in terms of rapidity or completeness of response is unknown and needs to be studied in a double-blind fashion. The only study that I am aware of that has compared the effectiveness of different doses of methotrexate in rheumatoid arthritis is one that was presented at this year's American Rheumatism Association Meeting in Minneapolis by *Russell* et al. [8] from Alberta Canada. They found no difference in effectiveness after 6 weeks of weekly i.m. methotrexate in 25 vs. 10 mg doses.

Another possible use of higher dose weekly intravenous methotrexate that should be looked into is whether it might be able to reinduce remission in patients whose disease has 'broken through' low-dose oral therapy.

A third question that is unanswered is whether weekly relatively-high-dose methotrexate is more toxic than low-dose oral methotrexate. Intuitively one would say yes, but in *Ward's* [9] placebo-controlled trial of low-dose oral pulse methotrexate in rheumatoid arthritis which was reported on this past June at the American Rheumatism Association Meetings, 29 patients on methotrexate and 11 on placebo were withdrawn because of adverse drug reactions, so that this question too would require a double-blind study. If high-dose methotrexate is no more effective than low-dose methotrexate, this becomes a moot point because there would be no reason to use high-dose methotrexate. On the other hand, if high-dose methotrexate is more effective than low-dose methotrexate, this becomes a very important question to answer because if the toxicity is significantly higher, then expected improvement must be significantly greater to warrant this increased risk.

Finally, in this day of escalating medical care costs, we should consider economics. In our area of Pennsylvania, the cost to the patient for 10 mg of oral methotrexate per week for 1 month is $ 12.45. The cost for 50 mg of intravenous methotrexate per week for one month is $ 75.12 and this does not include the added cost of office visits since the intravenous medication cannot be self-administered. One might also feel obligated to check laboratory tests more frequently if a high dose is used and again this would add to the cost.

So I feel then that methotrexate will be a useful drug in treating rheumatoid arthritis, but a lot more work needs to be done to determine the most effective way to give it.

References

1 Balin, Ph.; et al.: Is methotrexate therapy for psoriasis carcinogenic? A modified retrospective-prospective analysis. J. Am. med. Ass. *232:* 359–362 (1975).

2 Weinstein, G.D.: Methotrexate. Ann. intern. Med. *86:* 199–204 (1977).

3 Gubner, R.; et al.: Therapeutic suppression of tissue reactivity. II. Effect of aminopterin in rheumatoid arthritis and psoriasis. Am. J. med. Sci. *221:* 176 (1951).

4 Gross, D.; et al.: Die Immusupressive Behandlung der Progredient Chronischen Polyarthritis mit Antimetabolica und Cytostatica. Schweiz. med. Wschr. *97:* 1301–1310 (1967).

5 Michaels, R.M.; et al.: Weekly intravenous methotrexate in the treatment of rheumatoid arthritis. Arthritis Rheum. *25:* 339–341 (1982).

6 Willkens, R.F.; et al.: Low dose pulse methotrexate in rheumatoid arthritis. J. Rheum. *7:* 501–505 (1980).

7 Wilke, W.S.; et al.: Methotrexate in the treatment of rheumatoid arthritis. Pilot study. Clev. Clin. Q. *47:* 305–309 (1980).

8 Russell, A.S.; et al.: A double-blind placebo-controlled short-term trial of parenteral methotrexate in rheumatoid arthritis. 48th Ann. Meet. Am. Rheumatism Ass., Minneapolis 1984.

9 Ward, J.R., for the Cooperative Systematic Studies of Rheumatic Diseases: A prospective randomized controlled trial of low dose pulse methotrexate in rheumatoid arthritis. 48th Ann. Meet. Am. Rheumatism Ass., Minneapolis 1984.

Robert M. Michaels, MD, FACP, Chief, Department of Rheumatology,
Guthrie Clinic Ltd., Sayre, PA 18840 (USA)

Rheumatology, vol. 9, pp. 123–128 (Karger, Basel 1986)

Intravenous Treatment of Highly Active Rheumatoid Arthritis with Methotrexate

R. Rau, G. Herborn

Department of Rheumatology, Evangelisches Fachkrankenhaus, Ratingen, FRG

There are two publications concerning the efficacy and side effects of intravenous treatment of rheumatoid arthritis with methotrexate (MTX). One study was published by *Gross* et al. [1] in 1967, the other was published in 1982 by *Michaels* et al. [2]. Both of them could be invited to this meeting and just presented their papers.

As you will hear from Dr. *Karger,* we used to start MTX treatment always by intravenous application during the first 3 or 4 weeks but then switched to oral medication of MTX.

While *Gross* and *Michaels* treated their patients with doses of 50 mg once a week, our weekly dose was only 25 mg or even less.

To get an own impression of the short-therm effect of intravenous MTX and to contribute to todays discussion, we made a study in our department this year, comparable to the studies mentioned above: 10 consecutive patients, in whom MTX treatment was indicated, were not switched to oral medication after 3 or 4 weeks but stayed on intravenous application for a duration of 3–6 months.

Patients and Methods (table I)

Eight female and 2 male patients, age range from 29 to 73 years, mean age 49 years. Nine patients suffered from active rheumatoid arthritis, not sufficiently responding to gold or *D*-Penicillamine treatment. One patient suffered from psoriatic arthritis. One patient was in the morphological stage I, according to Steinbrocker, 4 patients were in stage II, 3 patients in stage III and 1 patient in stage IV.

Complete clinical and laboratory investigations were made at the beginning, after 1 month, after 2 months and at the end of the intravenous phase of MTX treatment. In

Table I. Patient data during intravenous treatment of 10 patients with MTX

Patient initials/age Sex/diagnosis and stage		Before MTX	After 1 month	After 2 months	Last control 3–6 months	Duration of treatment months	Total dose mg MTX	Weekly dose mg	Side effects
E.J. 32 years	JC	28	14		11	5	440	25	–
♀ RA II	SR	59	65		39				
	CO	5	0		0				
ML.B. 49 years	JC	37	57	14	13	5	360	15	nausea,
♀ RA II	SR	50	20	25	23				liver enzymes ↑
	CO	5	5	5	5				
E.L. 45 years	JC	34	22	11	32	3	225	15	liver enzymes ↑
♀ RA III	SR	70	78	52	59				
	CO	0	0	0	0				
W.M. 48 years	JC	45			1	5	585	25	–
♂ RA II	SR	124	80	51	12				
	CO	0	0	0	0				
H.G. 54 years	JC	40			32	4	305	25	stomatitis
♂ RA II	SR	20			31				
	CO	10			5				
M.G. 73 years	JC	42	15	14	20	3	90	5	nausea
♀ RA III	SR	60	60	67	60				
	CO	10	0	0	0				

D.B. 37 years	JC	79	37	29	27	3	325	25	–
♀ RA I	SR	141	57	75	70				
	CO	7.5	5	3.75	3.75				
J.E. 63 years	JC	30			8	4	280	15	nausea
♀ RA IV	SR	92	32	71	66				
	CO	5	3.75	3.75	3.75				
K.W. 61 years	JC	38	31	20	11	5	365	15	liver enzymes ↑
♀ RA III	SR	25	21	32	17				
	CO	20	0	0	0				
M.K. 29 years	JC	36	34	31	21	6	515	25	nausea, hair loss, liver enzymes ↑
♀ Pso.A.	SR	50	40	54	53				
	CO	0	10	7.5	0				
Mean value	JC	42	39	18	17				
49 years	SR	69	51	53	40.5				
	CO	6.25	2.8	1.6	1.7				

RA = Rheumatoid arthritis;
JC = joint count;
SR = sedimentation rate/1 h;
CO = corticosteroid dose per day (prednisone);
I—IV = anatomical stage of rheumatoid arthritis, according to Steinbrocker;
Pso.A. = psoriasis arthritis.

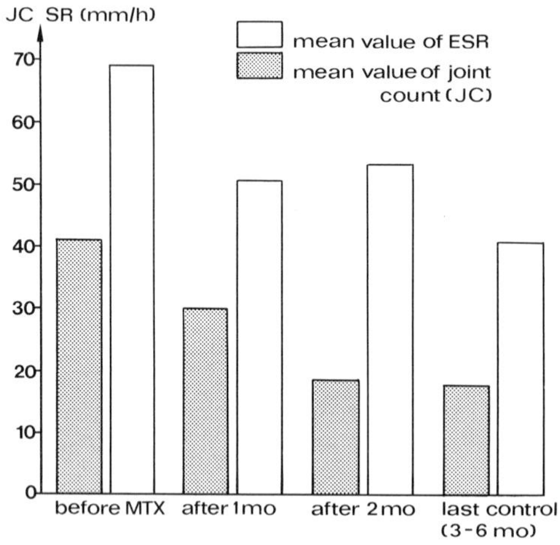

Fig. 1. Mean value of joint count and sedimentation rate before and during intravenous MTX treatment.

addition, during the first month weekly laboratory controls were made, including total blood count with thrombocytes, sedimentation rate, serum creatinine, SGOT, SGPT, γ-GT, alkaline phosphatase and urinalysis. After 1 month of treatment these laboratory investigations were made every 2 weeks.

To form the joint count all actively inflamed joints were counted. A joint with active synovitis – large and small joints – was counted as one point, a joint with active synovitis with effusion was counted as two points.

Results (table I)

At the beginning of treatment the joint count ranged from 28 to 79, the mean joint count was 42. The sedimentation rate ranged from 20 to 141 mm/h, mean 69 mm/h. It has to be taken in account that 7 of the patients were on oral corticosteroids, ranging from 5 to 20 mg of prednisone per day, mean 6.25 mg prednisone per day and that, in addition, in 4 patients a few intra-articular corticosteroid injections had to be given before starting MTX treatment. After starting MTX therapy no further intra-articular injections were allowed (fig. 1).

The first intravenous dose of MTX was 25 mg. If there were signs of intolerance, e.g. nausea or vomiting the following dose was reduced. So 5 patients stayed on 25 mg/week, in 4 patients the weekly dose was 15 mg and in 1 patient the weekly dose was 5 mg of MTX. At the end of the intravenous treatment after 3–6 months, the total dose of MTX was 90–558 mg, mean 349 mg.

Most of the patients already responded after 1 month of treatment: the joint count was reduced in 5 of 7 patients, in whom data are available. The mean joint count was 42 at the beginning of treatment and 39 after 1 month. The sedimentation rate was lower in 4 of 9 patients, the mean value dropped from 69 mm/h at the starting point to 51 mm/h after 1 month. This result was obtained in spite of the fact that the mean daily prednisone dose could be reduced from 6.25 to 2.8 mg.

At the end of treatment after 3–6 months, all but 2 patients experienced a very striking reduction of the joint count. The mean value dropped from 42 to 17. There was a clear-cut drop of the sedimentation rate in 6 patients, no significant change in 4 patients. But in 3 of these 4 patients the corticosteroid dose could be reduced. The mean value of the sedimentation rate dropped from 69 to 40.5 mm/h. The mean corticosteroid dose was reduced from 6.25 to 1.7 mg prednisone per day.

Side Effects

Three patients had no side effects. Three patients complained of nausea 6–24 h after the injections. In 4 patients slight elevations of SGOT and SGPT up to three times normal were detected.

One patient suffered from slight stomatitis and 1 had some loss of hair. These side effects did not cause a discontinuation of treatment. No bone marrow depression or infection was observed.

Discussion

In this open study 10 patients with highly active rheumatoid arthritis resistant to conventional forms of therapy were treated with intravenous methotrexate in weekly doses of 25 mg (in 5 patients), 15 mg (4 patients) and 5 mg (1 patient), respectively. After 3–6 months of treatment in 8 of 10 patients a striking and in 2 patients a moderate improvement of the joint count could be observed. In 6 patients there was an impressive drop of the sedimentation rate and in 6 of 7 patients who were on steroids, the steroid dose could be reduced. These results are comparable to the study of *Michaels* et al. [2], who used weekly doses of 50 mg MTX. Whether lower doses of intravenous MTX are as effective as higher doses has to be proven in a double blind manner. As *Michaels* mentioned, *Russel* et al. [3] did not find a difference in response between weekly intramuscular doses of 10 and 25 mg, respectively. Although *Michaels* did not find severe side reactions, in 4 of his 14 patients the dose schedule had to be changed or the therapy stopped because of intolerance. It has to be mentioned that *Gross* et al. [1], who treated their patients with 50 mg MTX per week and who obtained very good clinical results, did not continue to use MTX because the drug seemed to be too toxic for the treatment of rheumatoid arthritis. So, further studies have to be undertaken to compare different parenteral doses of MTX in respect of efficacy and toxicity to find the optimal weekly dose.

Conclusion

In conclusion, we can say that intravenous MTX treatment at doses between 15 and 25 mg is most effective in patients with very active rheumatoid arthritis not responding to conventional forms of therapy.

A clinical effect can be observed already after 4–6 weeks of treatment. During 3–6 months of MTX application only mild side effects can be seen, especially nausea and slight elevations of liver enzyme values. These side effects disappear with dose reduction or even with continuation of therapy. Further studies have to find the optimal weekly dose and to compare parenteral and oral forms of application.

References

1 Gross, D.; Enderlin; Fehr, K.: Die immunsuppressive Behandlung der progredient chronischen Polyarthritis mit Antimetaboliten und Zytostatika. Schweiz. med. Wschr. *97:* 1301 (1967).

2 Michaels, R.M.; et al.: Weekly intravenous methotrexate in the treatment of rheumatoid arthritis. Arthritis Rheum. *25:* 339 (1982).

3 Russel, A.S.; et al.: A double-blind placebo-controlled short-therm trial of parenteral methotrexate in rheumatoid arthritis. 48th Ann. Meet. Am. Rheumatism Ass., Minneapolis 1984.

PD Dr. med. R. Rau, Department of Rheumatology, Evangelisches Fachkrankenhaus, Rosenstrasse 2, D-4030 Ratingen (FRG)

Rheumatology, vol. 9, p. 129 (Karger, Basel 1986)

Discussion 5

Miescher: We will shortly discuss the three presentations. I remember, Dr. *Michaels* has mentioned the necessity for long-term treatment in order to maintain the effect of the methotrexate treatment, with the question, how should methotrexate be administered, what is the optimal dosage in comparison to toxicity, not to forget the cost problem, which incidently in comparison with cyclosporin would be twice as high. Who would like to comment on Dr. *Michaels* statements?

Rau: You told us that you could get a remission of the disease with 50 mg of MTX per week, but that the disease flared up again after reduction of the dose to 50 mg every 2 weeks. But, when 5–20 mg orally per week were given, then you could maintain the therapeutic effect. Does that mean that the oral dose is more effective than the intravenous application?

Michaels: We don't know. Our impression, and I emphasize impression, is that the patients we had in our study seemed to respond more completely and more, that's even hard to say more quickly than the people we have subsequently started on oral methotrexate. At this point in time, to be honest with you, we are not using intravenous methotrexate. Mainly, actually, because of the cost and inconvenience to the patient. At the time we did the study, as I said, we were just interested in 'does it work'? Because there really wasn't too much in the literature. As we were in the process of doing the study, in fact, Dr. *Wilke's* came out, and then their results seemed to be comparable to our results: giving orally is certainly more convenient to the patient, and less costly for the patient, so we subsequently went on to use low-dose oral once a week methotrexate. But again, our impression is that the people who receive the intravenous injections in a higher dose seem to get a more complete response on the lower dose of oral maintenance schedule.

Weinstein: There may actually be no advantage of intravenous over intramuscular. *Russell* and his colleagues published a paper on methotrexate levels in the Clinical Experimental Rheumatology in 1983, and they looked at methotrexate blood levels after either intravenous or intramuscular injection, and really there was virtually no difference. They looked at levels at 30 min, 1 h, 2 h, over a period of 24 h, and there was virtually no difference. So while there may be a difference between parenteral versus oral methotrexate, as far as bioavailability and serum levels, there is likewise no advantage of intravenous over intramuscular.

Miescher: Dr. *Michaels,* you might also consider to initiate remission with intramuscular or intravenous methotrexate and then to work out a suitable maintenance therapy?

Michaels: Yes, actually I have written a protocol for double-blind study. I just have to get people interested in it.

Miescher: I think the discussion of this topic will go on with Dr. *Willkens'* presentation.

Rheumatology, vol. 9, pp. 130–133 (Karger, Basel 1986)

Oral Methotrexate Therapy of Rheumatoid Arthritis

Robert F. Willkens

Harborview Medical Center, University of Washington, Seattle, Wash., USA

The use of methotrexate (MTX) to treat nonmalignant disorders occurred soon after its clinical availability by *Grubner and August* [1]. They used the drug to treat both psoriatic and rheumatoid disease. These early efforts were eclipsed by use of the drug in the treatment of neoplastic diseases.

The treatment of psoriasis provided the basis for low-dose use in non-neoplastic disease, and pioneered both intermittent and oral treatment regimens. *Van Scott et al.* [2] adapted intermittent oral pulse therapy of psoriasis with MTX to maximize the inhibition of the S-phase of the psoriatic epidermal cell. The drug was used orally by these investigators as MTX is completely absorbed in the gastrointestinal tract when given in doses of less than 30 mg/m^2 [2]. Clinical studies have since demonstrated an equimolar potency of low-dose oral and parenteral MTX [4]. Oral pulse regimens also permitted lower weekly doses, less comparable toxicity, lower blood levels, and were less expensive. Above all, the oral intermittent regimen was convenient for both patient and physician.

While many questions about the use of MTX to treat rheumatoid arthritis remained unanswered with respect to long-term toxicity, it seemed like a reasonable alternative to the disease-modifying therapeutic agents available a decade or more ago. At that time, neither penicillamine nor azathioprine were sanctioned for the treatment of RA and the oncogenic potential of the alkylating agents was already suspected [5].

Preliminary noncontrolled and anecdotal use were described using parenteral therapy. For example, *Hoffmeister* [6] treated 29 patients with 10–15 mg intramuscurlarly given once every 7 days. Two-thirds of the patients demonstrated decreased joint count, erythrocyte sedimentation rate and steroid requirement as a result of that therapy. *Wilke et al.* [7] in another pilot study examined the short-term benefit and toxic effects with 7.5 mg doses of MTX given weekly by mouth. After 3 months of therapy, the 6 participants of the study showed significant improvement of grip strength, joint count, and duration of morning stiffness.

We have used MTX over the last 15 years and continue to extend our original series which now encompasses an experience of over 150 rheumatoid patients. Some of these patients have been the subject of various earlier reports [8–10]. Prior to the study, most of the participants had been treated with nonsteroidal anti-inflammatory drugs, gold, *D*-penicillamine, and corticosteroids. Many had been treated with chloroquine and other immunosuppressive agents. In most patients, concomitant, nonsteroidal anti-inflammatory drugs and steroids were permitted; gold, penicillamine or immunosuppresives were disallowed. The process of patient selection was also guided by the exclusion of alcoholic, severely obese, and diabetic patients. Otherwise, healthy, elderly individuals were included if evidence of protein stores were judged to be adequate.

We used the intermittent pulse regimen developed by Weinstein, based on the psoriatic cell cycle [11]. Three oral 2.5–7.5 mg doses at 12-hour intervals are administered weekly. This regimen was used because it was associated with less short- and long-term side effects when compared with other regimens used to treat psoriatic patients. Patients were initially seen bimonthly and thereafter monthly for three successive months. Once established on the drug regimen, patients were seen every 12 weeks.

Patients judged the occurrence and severity of side effects and duration of morning stiffness. Subjective global assessment was graded as poor, fair or good. Joint tenderness and/or swelling was counted. Laboratory assessments included erythrocyte sedimentation rate, liver function tests and blood counts with the differentials.

Improvement by one or two global designations, that is, poor to fair or to good, occurred in over 85% of patients. Joint counts fell in most of those patients experiencing two global steps improvements. The ESR fell an average of 35 mm in those improving two global steps.

Among the earliest patients treated, it was our policy to discontinue therapy when liver enzyme determinations exceeded twice normal. Many patients on this regimen increased their liver enzyme values into a range of twice normal. In more recent use of MTX, therapy has not been interrupted upon the basis of this liver enzyme elevation inasmuch as liver biopsy of several such patients failed to reveal significant histopathology. In the event of persistent marked (three times normal) enzyme elevation, we have elected either to discontinue therapy or to perform a liver biopsy before continuing therapy.

More often therapy was discontinued for other reasons, such as inefficacy, gastrointestinal distress, pancytopenia, leukocytoclastic vasculitis [12], or apprehension. To date, 6 patients under our care have developed neoplasms while on MTX. One patient with pre-existing Sjogren's syndrome developed a non-Hodgkin's lymphoma which presented as a infiltrating shoulder mass. Three others have developed adenocarcinomas and one a malignant melanoma. One patient developed a squamous cell carcinoma of the skin.

The occurrence of pancytopenia has been related for the most part to the coexistence of renal disease, either pre-existing or induced by concomitant disease or drugs [13].

Pulmonary toxicity as described by *Cannon et al.* [14] was not seen among these patients. We did, however, encounter minor side effects such as hypersensitivity [15].

Conclusion

Improvement of 85% of 150 patients treated with oral, intermittent, low-dose MTX has confirmed the impression of the effectiveness of this regimen. Patient selection must anticipate the enhanced risk in patients where there may be concomitant liver or renal toxicity. Those patients are also at increased risk who present with depleted protein and folate stores. There is probably a liver biopsy in the natural history of the rheumatoid patient taking MTX, but when and how often needs yet to be determined. If this treatment is immune modulating, the occurrence of a non-Hodgkin's lymphoma and two skin neoplasms as demonstrated by these patients must be considered as possible adverse effects.

References

1 Grubner, R.S.; August, V.R.: Therapeutic suppression of tissue reactivity. II. Effect of aminopterin in rheumatoid arthritis and psoriasis. Am. J. med. Sci. *220:* 176 (1950).
2 Van Scott, E.F.; Auerbach, R.; Weinstein, G.D.: Parenteral methotrexate in psoriasis. Archs Derm. *89:* 550 (1964).
3 Huffman, D.H.; Wan, S.H.; Aranoff, D.L.; et al.: Pharmacokinetics of methotrexate. Clin. Pharm. Ther. *14:* 572 (1973).
4 Acute Leukemia Group B; Acute lymphocytic leukemia in children. Maintenance therapy with methotrexate administration intermittently. J. Am. med. Ass. *207:* 923 (1969).
5 Cooperating Clinics Committee of the American Rheumatism Association: A controlled trial of cyclophosphamide in rheumatoid arthiritis. New Engl. J. Med. *283:* 883 (1970).
6 Hoffmeister, R.T.: Methotrexate in rheumatoid arthritis (Abstract). Arthritis Rheum. *15:* 114 (1972).
7 Wilke, W.S.; Calabrese, L.H.; Scherbel, A.L.: Methotrexate in the treatment of rheumatoid arthritis. Cleve. Clin. Q. *47:* 305 (1980).
8 Willkens, R.F.; Watson, M.A.; Paxson, C.P.: Low dose pulse methotrexate therapy in rheumatoid arthritis. J. Rheum. *7:* 501 (1980).
9 Willkens, R.F.; Watson, M.A.: Methotrexate: a perspective of its use in the treatment of rheumatic diseases. J. Lab. clin. Med. *100:* 314 (1982).
10 Willkens, R.F.; Williams, H.S.; Ward, J.R.; et al.: Randomized, double-blind, placebo controlled trial of low-dose pulse methotrexate in psoriatic arthritis. Arthritis Rheum. *27:* 945 (1984).
11 Weinstein, G.D.; Frost, P.: Methotrexate for psoriasis. Archs Derm. *103:* 33 (1971).
12 Marks, C.; Willkens, R.F.; Wilske, K.R.; Brown, P.B.: Small vessel vasculitis and methotrexate. Ann. intern. Med. *100:* 916 (1984).
13 MacKinnon, S.K.; Starkebaum, G.; Willkens, R.F.: Pancytopenia associated with the use of low-dose methotrexate in the treatment of rheumatoid arthritis. Semin. Arthritis Rheum, (in press).
14 Cannon, G.W.; Ward, J.R.; Clegg, D.O.; Samuelson, S.; Abbott, T.A.: Acute lung disease associated with low dose pulse methotrexate treated patients with rheumatoid arthritis. Arthritis Rheum. *26:* 1269 (1983).
15 Weiss, R.B.; Bruno S.: Hypersensitivity reaction to cancer chemotherapeutic agents. Ann. intern. Med. *94:* 66 (1981).

Dr. Robert F. Willkens, Harborview Medical Center, University of Washington, 325 Ninth Avenue, Seattle, WA 98104 (USA)

A H

Rheumatology, vol. 9, pp. 134–144 (Karger, Basel 1986)

Incidence of Toxicity in Patients with Rheumatoid Arthritis Treated with Methotrexate

William S. Wilke, Leonard H. Calabrese, Patricia L. Krall, Allen M. Segal

Cleveland Clinic Foundation, Cleveland, Ohio, USA

Introduction

Numerous open studies [1–6] and 2 new controlled studies [7, 8] have suggested that low-dose, pulse methotrexate can be used with some success as treatment for patients with resistant rheumatoid arthritis. As with all cytotoxic drugs, toxicity is a major concern. This paper is a retrospective analysis of the long-term toxicity of methotrexate in patients with resistant rheumatoid arthritis. Efficacy is also estimated.

Materials and Methods

The medical records of 87 unselected patients treated at the Cleveland Clinic with the diagnosis of seropositive definite or classical rheumatoid arthritis who were treated with methotrexate were reviewed to determine the type and incidence of untoward effects. Methotrexate had been initiated between the years of 1976 and 1982 in 78 patients who were followed solely by the authors. Methotrexate had been initiated before 1976 in 9 other patients who had been followed in part by other physicians at the Cleveland Clinic.

All patients had either failed to respond to or were unable to tolerate more conservative therapy including anti-inflammatory doses of aspirin or nonsteroidal anti-inflammatory drugs, moderate dose corticosteroid, hydroxychloroquine and in some cases parenteral gold, *d*-penicillamine and other cytotoxic drugs. All patients were judged to have active disease as evidenced by elevated acute phase reactants and by physical examination showing 5 or more active joints.

All concomitant drugs were recorded. Also recorded was the presence of diabetes and obesity in this patient population. Diabetes was defined as a 2-hour glucose tolerance test in which both 1- and 2-hour values exceeded 200 mg% or a 3-hour value which exceeded 140 mg%. Obesity was defined as weight 30% greater than the optimum for a given height defined on a recognized nomogram [9].

When the addition of methotrexate was the only change in patients' therapy and disease parameters including active joint count, grip strength, duration of morning stiffness and Westergren sedimentation rate were properly recorded at the initiation of methotrexate and after approximately 3 months of treatment, efficacy was estimated. Quantitation for at least 3 of 4 measurements were required at baseline and 3 months. The percentage of improvement from baseline was averaged.

Results

The mean cumulative dose of methotrexate for all 87 patients was 1,397 mg (range 45–3,875 mg), the mean weekly dose 7.84 mg (range 2.5–12.5 mg) and the mean duration of therapy 42.1 months (range 1.5–126 months). Sixty-three patients apparently suffered no untoward effects considered to be related to methotrexate. This patient population was composed of 39 females and 24 males with a mean age of 51.27 years (range 26–76 years). The mean cumulative dose of methotrexate in this group was 1,238 mg (range 45–3,820 mg), mean weekly dose 7.74 mg (range 2.5–12.5 mg) and the mean duration of therapy was 40.02 months (range 1.5–120 months). Twenty-four patients (26%) experienced signs and symptoms which might have been side effects or toxicity from methotrexate. This patient population was composed of 15 females and 9 males whose mean age was 55.6 years (range 32–69 years). The mean cumulative dose of methotrexate in this group was somewhat higher at 1,816 mg (range 90–3,875 mg), mean weekly dose 8.11 mg (range 7.5–12.5 mg) and mean duration of therapy 47.71 months (range 3–126 months).

Twenty-seven episodes of possible toxicity occurred in 24 patients. These episodes included elevation of liver enzymes in 13, peptic ulcer 3, nausea 2, mouth ulcers 2, alopecia 2, herpes zoster 2, concurrent neoplasm 2 and overdose with mucositis 1. The details are included in table 1. Thirteen percent of the patients who apparently suffered no toxicity were obese compared to 12% of the patients who experienced toxicity. Diabetes mellitus was not seen in the population who experienced toxicity. Five patients with maturity onset diabetes controlled by diet and 1 patient with insulin dependent diabetes mellitus were found in the population who experienced no toxicity.

Table I. Toxicity in 24/87 patients

Number of occurrences	Mean cumulative dose, mg (range)	Mean weekly dose, mg	Untoward reactions
13	1,058 (90–3,722)	7.85	SGOT or SGPT elevated
3	733 (180–1,780)	7.69	peptic ulcer – all taking NSAID's
2	710, 30	8.40	nausea[1]
2	990, 210	7.50	mouth ulcers
2	180, 60	7.50	alopecia
2	510, 2,340	7.60	herpes zoster
2	585, 540	7.50	neoplasm adenocarcinoma rectum[1] metastatic adenocarcinoma[1]
1	–	100 mg in one week	overdose, mucositis

[1] Methotrexate discontinued in patients.

Table II. Details of hepatic toxicity in 13 patients

Patient no.	ETOH	Cumulative dose, mg	Weekly dose, mg	SGOT, μ/l (0–40)	SGPT, μ/l (0–30)	Liver biopsy
Isolated elevation of enzymes						
21	no	2,160	7.5	52	–	normal
1	yes	180	7.5	41	–	–
2	yes	90	6.0	45	–	–
14	yes	570	7.5	–	35	–
15	no	660	7.5	45	–	–
17	yes	360	7.5	70	–	–
4	no	360	7.5	42	–	–
Two or more elevations						
11	no	3,722	7.69	70.57	–	portal fibrosis[1]
12	yes	2,400	12.5	55.225	–	normal + triaditis[1]
13	no	1,224	8.5	72.65	–	rare necrotic cells[1]
16	yes	90	7.5	43.52	–	–
18	yes	0	7.5	–	36.45	–
3	yes	0	7.5	48.60	–	–

[1] Methotrexate discontinued.

Table III. Concurrent medications

	Number of patients	Hydroxy-chloroquine		NSAID's		Prednisone		Folic acid	
		n	%	n	%	n	%	n	%
No toxicity	63	45	71	62	98	48	76	50	79
Toxicity	24	20	83	21	88	20	83	20	83

Methotrexate was discontinued because of toxicity in 6 patients, 3 because of episodic elevation of SGOT and/or SGPT enzymes, 1 for intractable nausea and 2 after the appearance of adenocarcinoma of the rectum and metastatic adenocarcinoma, respectively. The patient with nausea underwent extensive testing which failed to demonstrate pathology. He subsequently experienced more severe nausea while taking azathioprine and is currently again taking methotrexate without gastrointestinal complaints.

In table II we have documented the details of hepatic toxicity in 13 patients. In 8 of the 13 patients the use of alcohol appeared to be a complicating factor. There did not appear to be a trend toward a higher incidence of enzyme elevation at higher cumulative doses since elevated enzymes could occur at any cumulative dose. Five liver biopsies were obtained in 4 patients. These biopsies were usually necessitated by elevated enzymes or high cumulative dose of methotrexate. Portal fibrosis was seen in 1 patient who had also had enzyme elevations. Methotrexate was discontinued. This patient had taken a cumulative dose of 3,722 mg of methotrexate. She denied the use of alcohol. No other significant abnormalities were seen in the other liver biopsies.

The percentage of patients taking concurrent hydroxychloroquine, prednisone, nonsteroidal anti-inflammatory drugs and folic acid are presented in table III. We specifically looked at the influence which hydroxychloroquine might have on methotrexate toxicity. Of the patients who experienced toxicity, 20 of 24 patients were receiving hydroxychloroquine. Of the 63 patients who experienced no toxicity 45 were receiving hydroxychloroquine and 18 were not. Although a slightly higher percentage of patients who experienced toxicity were taking plaquenil, all patients who experienced toxicity and all who

Table IV. Plaquenil and toxicity

Patient group	Toxic		Not toxic		
	n	%	n	%	
Methotrexate and plaquenil	20	23	45	52	$p = 0.25$[1]
Methotrexate alone	4	5	18	21	

[1] χ^2.

Table V. Plaquenil, dosage and toxicity

Patient group	Number of patients	Mean cumulative dose, mg%		Mean weekly dose mg%	
Methotrexate and plaquenil	65	1,522	$p = 0.07$*	7.92	$p = 0.33$*
Methotrexate alone	22	1,026		7.62	
Toxicity	24	1,816	$p = 0.16$*	8.11	$p = 0.93$*
No toxicity	63	1,238		7.74	

* T test.

were taking plaquenil tended to have received a higher mean cumulative dose of methotrexate and tended to be taking a somewhat higher weekly dose of the antimetabolite (table IV, V). However, the magnitude of the cumulative dose, weekly dose or the addition of hydroxychloroquine did not influence the incidence of toxicity. We feel that the addition of hydroxychloroquine did not increase the incidence of toxicity.

Efficacy was estimated in 26 patients (table VI). Of the 87 patients studied for toxicity, only the charts of these 26 patients were complete enough to be used to provide a rough estimate of overall response. Eighty-five percent of patients showed an overall improvement of 30% or better from baseline. Measurements of morning stiffness and the joint count were more likely to improve than were measurements of Westergren sedimentation rate and grip strength.

Table VI. Efficacy of methotrexate in 26 patients, % improvement[1]

< 0	0	
0–9	1 (4%)	
10–19	2 (8%)	
20–29	1 (4%)	
30–39	1 (4%)	85%
40–49	1 (4%)	
50–59	4 (15%)	
60–69	3 (12%)	
70–79	9 (35%)	
80–89	4 (15%)	
90–100	0	

[1] Based on average of at least 3 of 4 parameters including morning stiffness, grip strength, joint count and Westergren sedimentation rate.

Discussion

It is difficult to compare the incidence of toxicity of methotrexate in our study with other published studies because of the retrospective nature of this study and the length of the period of follow-up (table VII). Only *Hoffmeister's* [6] 1983 publication lends itself to this kind of comparison. *Hoffmeister* used 10–15 mg of parenteral methotrexate given for a mean of 4.68 years to 78 patients and noted a 17% incidence of significant toxicity. In other studies such as *Willkens and Watson* [2] in which similar doses of methotrexate were used the incidence of significant side effects was estimated at 13%. Even though similar doses of methotrexate were used in the prospective studies of *Ward* [7] and *Russell* et al. [8] minor toxicity appears to be higher than in our study. However, it is clear that when higher weekly doses of methotrexate are used such as in the separate reports by *Steinsson* et al. [4] and *Michaels* et al. [5], the incidence of significant toxicity climbs. It is well to remember that the majority of the toxicity which we have observed can be readily controlled by minor changes in the dosage. However, because we do feel that the incidence of toxicity will roughly correlate with the magnitude of the weekly dose of methotrexate, we suggest that the lowest possible dose be used, between 5 and 12.5 mg/week, and that this dose be given as 1 dose. If gastrointestinal intolerance is a problem the dose can be divided so that the entire dose is given within 12 h.

Table VII. Methotrexate toxicity in the medical literature

Investigator year	Number of patients	Weekly dosage, mg duration	Percent toxicity	Elevated liver function tests	Nausea	Stomatitis	Other
Hoffmeister [6], 1983	78	10–15 4–68 years	17	8% (6)		8% (6)	3% (2)
Willkens et al. [2], 1980	32	7.5–15 1–10 years	13 'significant'	6% (2)	'many'	–	–
Ward [7], 1984	95	7.5–15 18 weeks	30.5	20% (18)	–	–	4% (4)
Groff et al. [3], 1982	28	5–12.5 12.5 months	50	14% (4)	18% (5)	11% (3)	25% (7)
Steinsson et al. [4], 1982	21	7.5–25 38 weeks	57	28% (10)	14% (3)	14% (3)	29% (6)
Michaels et al. [5], 1982	14	10–50 7–20 weeks	86	36% (5)	64% (9)	21% (3)	0

We also employ the use of folic acid as another theoretical measure to reduce toxicity. This is generally given in equal mg approximately 5 days after the methotrexate is given. From the separate works of *Werkheiser* [10] and *White* [11] folic acid might provide a direct protective effect by displacing the drug from the enzyme dihydrofolate reductase. This is highly controversial. More simply, by giving weekly folate we may be preventing patients from developing subclinical folate deficiency. It is well known from the work of *Hellman* et al. [12] and others [13] that the incidence and severity of methotrexate toxicity is increased in patients who are folate deficient.

Although some authors feel that the concurrent use of hydroxychloroquine and cytotoxic drugs may be dangerous because of the property of chloroquine to inhibit DNA repair and slow the rate of protein synthesis in lymphoblastic leukemia cells [14], we do not feel that the addition of hydroxychloroquine confers a greater incidence of toxicity. In fact, we feel that the use of other medications which might include nonsteroidal anti-inflammatory drugs, low-dose corticosteroids and hydroxychloroquine might be beneficial. Present knowledge about the pathogenesis of rheumatoid arthritis suggests that immune

Table VIII. Incidence of toxicity with other slow-acting agents

Agent author	Number of patients	Pro-spective	Retro-spective	Duration of follow-up	Significant toxicity
d-Penicillamine *Rothermich* et al. [16]	200		×	23–62 months	48% (96)
Cyclophosphamide Cooperating Clinics [17]	48	×		32 weeks	40% (13) LD 90% (18) HD
Azathioprine *Hunter* et al. [18]	12	×		40 months	33% (4)
Gold *Rothermich* et al. [19]	97	×		4 years	29% (28)

LD = 5–15 mg/day; HD = 50–150 mg/day.

mechanisms initiate the inflammatory cascade which then results in synovial proliferation, pannus formation and joint destruction. These chemical and cellular events are ordered, one preceding and amplifying the next. Each individual drug used to treat the disease attenuates 1 or more of these events. Complete suppression of any individual event with 1 agent is difficult unless potentially toxic quantities of this agent are used. Because amplification occurs in the pathogenic mechanisms which result in joint destruction, attenuation at multiple sites in this sequence may result in drug potentiation.

Recently *McCarty's* group [15] reported good response in 14 of 17 patients with resistant rheumatoid arthritis who were treated with a combination of low-dose cyclophosphamide, azathioprine and hydroxychloroquine for a mean of 7 months. This combination also had a corticosteroid-sparing effect. *McCarty* and associates have demonstrated that a polypharmacologic approach allows the use of multiple drugs at low dose presumably in order to minimize individual toxicity. We suspect that we are able to employ slightly lower weekly doses of methotrexate thereby increasing the efficacy of low-dose drug and reducing its potential toxicity for the same reasons.

Weinstein [15] has suggested that patients who are obese or have diabetes mellitus are at a greater risk of liver toxicity from the long-term use of methotrexate. We do not have the liver biopsy data to address this question and can not specifically comment about that relationship. However, neither obesity nor diabetes predicted toxicity in this relatively long-term study. The psoriasis literature [15] has also suggested that cumulative dose varies directly with toxicity. We were unable to confirm this relationship.

How does methotrexate compare with other slow-acting or disease modifying antirheumatic drugs? Again it is difficult to compare our data with toxicity data from most of the available studies (table VIII). The incidence of toxicity with methotrexate certainly compares favorably to that of *d*-penicillamine [16] and cyclophosphamide [17]. *Hunter* et al. [18] reported an incidence of toxicity of 33% in 12 patients receiving azathioprine for a comparable follow-up period of 40 months. *Rothermich* et al. [19] reported an incidence of 29% toxicity in a 4-year prospective study of parenteral gold. The expected toxicity from methotrexate would appear to parallel these last 2 studies.

Although acute and chronic effects from all of these drugs are worrisome, of even greater concern is the possibility of future malignancy. While methotrexate is assumed to be teratogenic, it does not appear to be oncogenic [20]. Unfortunately, oncogenicity is well documented for alkylating agents [21] and may occur with azathioprine [22]. The consideration of oncogenicity alone certainly favors methotrexate as a potentially safe cytotoxic agent. In addition, the onset of objective response to methotrexate appears to occur earlier than with other cytotoxic agents [3] generally within 4–8 weeks. We believe that the apparent lack of oncogenicity, acceptable and mild dose-related acute toxicity and relatively faster onset of disease modification may also favor methotrexate as the cytotoxic agent for the treatment of resistant rheumatoid arthritis.

References

1 Wilke, W.S.; Calabrese, L.H.; Scherbel, A.L.: Methotrexate in the treatment of rheumatoid arthritis. Pilot study. Cleve. clin. Q. *47:* 305–309 (1980).
2 Willkens, R.F.; Watson, M.A.: Methotrexate. A perspective of its use in the treatment of rheumatic disease. J. Lab. clin. Med. *100:* 314–321 (1982).

3 Groff, G.D.; Shenberger, K.N.; Wilke, W.S.; Taylor, T.H.: Low-dose oral methotrexate in rheumatoid arthritis. An uncontrolled trial and review of the literature. Semin. Arthrit. Rheumatism *12:* 333–347 (1983).

4 Steinsson, K.; Weinstein, A.; Korn, J.; Ables, M.: Low-dose methotrexate in rheumatoid arthritis. J. Rheum. *9:* 860–866 (1982).

5 Michaels, R.M.; Nashel, D.J.; Leonard, a.; Sliwinski, J.; Derbes, S.J.: Weekly intravenous methotrexate in the treatment of rheumatoid arthritis. Arthritis Rheum. *25:* 339–341 (1982).

6 Hoffmeister, R.T.: Methotrexate in rheumatoid arthritis. 15 years experience. Am. J. Med. Oral Gold Symposium: 69–73 (1983).

7 Ward, J.R.: For the cooperative systematic studies of rheumatic disease: A prospective randomized controlled trial of low-dose pulse methotrexate in rheumatoid arthritis. Arthritis Rheum. *27:* suppl., p. S40 (1984).

8 Russell, A.S.; Watts, C.; Thompson, R.; Edelman, J.; Esdaile, J.; Percy, J.; Davis, P.; Tannenbaum, H.; Grace, M.: A double-blind placebo-controlled short-term trial trial of parenteral methotrexate in rheumatoid arthritis. Arthritis Rheum. *27:* suppl. p. S59 (1984).

9 Bray, G.A.: Definition, measurement and classification of the syndromes of obesity. Int. J. Obesity *2:* 99–112 (1978).

10 Werkheiser, W.C.: Specific finding of 4-amino folic acid analogs by folic acid reductase. J. biol. Chem. *236:* 888–893 (1961).

11 White, J.C.: Reversal of methotrexate binding to dihydrofolate reductase by dihydrofolate. J. biol. Chem. *254:* 10889–10894 (1979).

12 Hellman, S.; Ianotti, A.T.; Bertino, J.R.: Serum folate activity in patients with tumors of the head and neck treated with methotrexate. Proc. Am. Ass. Cancer Res. *4:* 27 (1963).

13 Ramakrishnan, S.; Houston, L.L.: Inhibition of human acute lymphoblastic leukemia cells by immunotoxins. Potentiation by chloroquine. Science *223:* 58–61 (1984).

14 McCarty, D.J.; Carrera, G.F.: Intractable rheumatoid arthritis. Treatment with combined cyclophosphamide, azathioprine and hydroxychloroquine. J. Am. med. Ass. *248:* 1718–1723 (1982).

15 Weinstein, G.; The Writing and Organizing Committee for the Cooperative Study: Psoriasis-liver-methotrexate interactions. Archs Derm. *108:* 36–42 (1973).

16 Rothermich, N.O.; Thomas, M.H.; Phillips, V.K.; Bergen, W.: Clinical trial of penicillamine in rheumatoid arthritis. Arthritis Rheum. *24:* 1473–1478 (1981).

17 Cooperating Clinics Committee of the American Rheumatism Association: A controlled trial of cyclophosphamide in rheumatoid arthritis. New Engl. J. Med. *238:* 883–889 (1970).

18 Hunter, T.; Urowitz, M.B.; Gordon, D.A.; Smythe, H.A.; Ogryzlo, M.A.: Azathioprine in rheumatoid arthritis. A long-term follow-up study. Arthritis Rheum. *18:* 15–20 (1975).

19 Rothermich, N.O.; Phillips, V.K.; Bergen, W.; Thomas, M.H.: Crysotherapy. A prospective study. Arthritis Rheum. *19:* 1321–1327 (1976).

20 Bailin, P.L.; Tindall, J.P.; Roenigk, H.H., Jr.; Hogan, M.D.: Is methotrexate therapy for psoriasis carcinogenic? A modified retrospective-prospective analysis. J. Am. med. Ass. *232:* 359–362 (1975).

21 Baltus, J.A.M.; Boersma, J.W.; Hartman, A.P.; Vandenbroucke, J.P.: The occurrence of malignancies in patients with rheumatoid arthritis treated with cyclophosphamide. A controlled retrospective follow-up. Ann. rheum. Dis. *42:* 368–373 (1983).

22 Grunwald, H.W.; Rosner, F.: Acute leukemia in immunosuppressive drug use. A review of patients undergoing immunosuppressive therapy for non-neoplastic diseases. Archs intern. Med. *139:* 461–466 (1979).

William S. Wilke, MD, Cleveland Clinic Foundation, 9500 Euclid Avenue, Cleveland, OH 44106 (USA)

Rheumatology, vol. 9, pp. 145–147 (Karger, Basel 1986)

Discussion 6

Miescher: Thank you very much, Dr. *Wilke,* for the beautiful presentation. We are now discussing Dr. *Willkens'* and Dr. *Wilke's* presentation. Just to mention one point. If you are doing this alternate placebo-methotrexate study, you have of course also seasonal influences. For instance, if you start in a season where people have a lot of intercurrent viral conditions, you may have a high C19-binding a high sedimentation rate. If you approach spring, when all those diseases are fading away, you get an improvement which appears to be a placebo improvement, but which is nothing else than a seasonal development.

Wilke: I agree entirely with you. That would be very statistically hard to deal with. So, I can't tell you which patients started in early spring, which is definitely the time to start anyone on a study. I can't give you that information, I haven't worked out, but what you say is obviously correct.

Rau: One word to your study, Dr. *Willkens,* I think that 7.5 mg per week is a very low dose to compare with placebo. In our patients, we began with much higher doses and then tried to reduce the dose as much as possible. And there are only a few patients in whom we could reduce the dose below 12.5 or 10 mg. So I think 7.5 mg is not a dose with which you can get a most impressive difference to placebo.

Willkens: I think you will find, when Dr. *Williams* from the co-operative clinics, that the people who responded to 7.5 mg do so in an exemplary fashion and probably those that do respond are a subset that do very well. And that's why I'm glad we have that data. Seven and a half is an effective dose.

Rau: But not in all patients and not in a very much convincing way. Now some comments perhaps to your presentation, Dr. *Willkens.* First, we have the same impression as you, that the number of nodules and the size of nodules seem not to be reduced by MTX. I don't know if there are more nodules, but there are not fewer. And secondly, as I understood, you got the severe side effects only in patients who had at the same time salicylates or who had renal failure. And can you tell me the creatinine level in those two patients with cytopenia?

Willkens: Yes, in one patient it was over 2, 2.3; and in the person that got the leukocytoclastic angiitis it was in the order of 4. And as I remember, it wasn't salicylates, it was another nonsteroidal agent that the patient received.

Miescher: Dr. *Baken* in Birmingham responded to the nodule thing by suggesting that we might be converting somebody with an exudative to a tuberculoid form of rheumatoid arthritis, which is a very interesting hypothesis. That we do something to the immune apparatus and we change the nature of the disease, and most of us feel that we are not inducing a remission, we are inducing something that will alter the state of inflammation, maybe the immune system. And in so doing, we may be allowing some other manifestation of the disease to become more prominent. Now, let us first discuss the topic of side reactions.

Weinstein: I have a question for Dr. *Willkens:* you imply, and I don't know if you mean it, that solar dermatitis was the same as your cutaneous leukocytoclastic angiitis. I don't know whether you biopsied those 2 patients.

Willkens: We didn't biopsy the regular solar dermatitis... I don't know if it's the same, I didn't think so.

Weinstein: Because the solar dermatitis is described in the literature and in biopsy is really quite different than the leukocytoclastic angiitis.

Willkens: No, the answer is no.

Rau: Have you done liver biopsies before treatment or only after treatment?

Willkens: No pretreatment biopsies were done.

Rau: So, you have to keep in mind that rheumatoid arthritis per se can be associated with liver disease. In former biopsy studies we found a mild to moderate reactive hepatitis in about one-third of cases even in early cases of RA without any treatment. So I think, if you wish to evaluate the effect of MTX, you would have to do series of liver biopsies to see the development.

Willkens: Well, I'm not sure that earlier data on hepatic disease in arthritis is quite accurate. And I'm not prepared to accept the fact that there is that much pre-existing liver disease in rheumatoid patients. There is a lot of data about hepatocellular liver function tests and people who have rheumatoid disease. I'm not sure that this relates to rheumatoid disease. And there are specific cellular changes that occur in some patients, where there is a correlation. But, our pathologists really question that rheumatoid disease itself produces hepatic disease, and I'm sorry that Dr. *Mackenzie* isn't here, because he did a double-blind study in the days when the patients would allow you to do a biopsy in normal livers, as were in patients with rheumatoid arthritis. And as I recall, these date do not support the fact that rheumatoid arthritis predisposes to intrinsic hepatocellular disease itself. After all, how many patients do we see who are not on some form of therapy.

Rau: You know, I read that paper by *Mackenzie.* He found no difference between patients with treatment or under treatment with MTX and patients without treatment. And in those patiens without treatment he diagnosed a reactive hepatitis in 20% of cases. And I think we shall discuss the problem this afternoon.

Wilke: In a superficial way I will present that data this afternoon and then we wrestle with it...

Miescher: Maybe also in the time we got, when biopsies were done on normal people, a study was done on American stewardesses, and they had a surprisingly high incidence of mild liver damage.

Karger: A question to Dr. *Willkens.* You were talking about the leukocytoclastic angiitis. Did you see it in your first experience with MTX or in the later years, when your interest was stuck upon this illness by dermatologists? And a second question, did you withhold MTX therapy in these cases or continue?

Willkens: Those are good questions. First of all, this came in about the 15th year of use of methotrexate, and the question is, why hadn't I seen it before? I think the thing that happened is, I saw one and biopsied it and said: oh, that's what that is, whereas previously one had disregarded it, or been able to talk it away, or... I just hadn't seen it prior to that time. But once you've seen one, it's almost as though you see, you begin to recognize what you are seeing as being something more definitive. So I think, it probably existed, I just didn't pay any attention to it. We did discontinue methotrexate in the majority of those patients and we challenged them all subsequently. And something like two-thirds of them again got the rash. So, occasionally, it is possible to start it and not get the rash again, and because it took so long for the rash to develop, I'm not sure that those people who didn't get it right away wouldn't get it later again. It probably relates to factors that we don't understand, like some sort of cumulative doses, as well as maybe sum exposure, perhaps concomitant medication, I don't know.

Wilke: I just want to make one comment because Dr. *Willkens* has brought up an interesting point we'll probably hear more about, and that is pulmonary toxicity. Because of Dr. *Williams,* we specifically went through these trials again, looking for episodes of pneumonitis, etc. And we found four episodes of pneumonitis in this group of patiens, all very typical bacterial pneumonitis that respond not to the stopping of methotrexate or anything else but responded very quickly to antibiotics. And so, no pulmonary toxicity that we would call methotrexate-induced in any of our patients, which I thought was interesting, because, we've heard a statistical 5% of incidence.

Rau: One question to Dr. *Willkens.* Do you think that the rate of side effects was dose dependent? And a second question: Do you think that there is any difference in toxicity between the divided dose and the single weekly dose?

Willkens: Well, the data that pre-existed in the area of psoriasis, not withstanding what *Nyfors* said yesterday, I consider to demonstrate that the intermittent dose is less toxic. I mean, the difference in toxicity between the single dose and the divided dose really wasn't very great. And, I think at these dosage levels there isn't a lot of difference. I don't think that there is good evidence anywhere to support the contention that the divided dose is more toxic. If anything, the data that existed prior to this time show that it's less toxic.

Rau: So the position of *Nyfors* is quite contrary. In his opinion, with respect to liver toxicity, the divided dose is more toxic than the single dose.

Willkens: I haven't seen that information. And as a matter of fact, there are contrary data from *Weinstein,* which suggest that the divided dose is less toxic.

Williams: I would have to say I agree with Dr. *Willkens* on that. After recently reviewing it, for the publication of our data, that all the data would indicate thate it is less toxic.

Miescher: I think we have to continue the discussion during coffee. I just want you to remember that we are at the stage, where oncologists have been some 20 years ago, to face the problem of efficacy versus toxicity. And we have to find out ways and means to provide better care for the patiens by possibly combining, either simultaneously or in alternation, drugs, in order to maintain long-lasting remissions with a minimum amount of toxicity. It took the oncologists about 15 years to make significant progress, we may discuss this during the coffee break.

Rheumatology, vol. 9, pp. 148–153 (Karger, Basel 1986)

Effect of Methotrexate on Rheumatoid Arthritis: A Systematic Analysis of 21 Patients[1]

Arthur Weinstein, Sally Marlowe, Joseph Korn, Faripour Farouhar

Division of Rheumatic Diseases, Department of Medicine and Department of Pathology, University of Connecticut Health Center and VA Medical Center, Farmington, Conn., USA

In 1982 we reported on the short-term effectiveness and toxicity of methotrexate (MTX) in 21 patients with rheumatoid arthritis (RA) [1]. The purpose of the present study was to reevaluate the original group of patients to determine: (1) whether MTX continued to be clinically effective; (2) whether serious adverse reactions occurred, and (3) whether serious liver toxicity appeared.

Patients and Methods

The 21 patients had severe classical RA with active disease and had received prior treatment with gold and at least one other second line agent [1]. There were 15 females and 6 males with a mean age of 51 years and a mean duration of RA of 9.4 years. MTX was instituted at a dose of 7.5–15 mg once weekly, usually in 2 or 3 divided doses. Initially, 18 patients received oral MTX and 3 intramuscular. At the first report (follow-up I), 3 patients had discontinued MTX. Subsequently, 1 of these patients has been restarted on MTX and 4 others had discontinued treatment. Patients have had complete blood count, platelet count and serum transaminases every 4–8 weeks.

For this study (follow-up II), the 15 patients remaining on MTX were assessed for disease activity by determination of the number of tender and swollen joints, duration of morning stiffness, mean grip strength, global assessment of improvement and sedimentation rate.

Liver biopsies were done in 13 of the 15 patients. In addition, 4 other RA patients followed by us had liver biopsies after having received at least 1 g MTX and are included in the analysis of liver toxicity. Biopsy specimens were stained with hematoxylin and eosin, PAS, PAS-diastase and trichrome to examine for steatosis, variation in nuclear size, nuclear vacuolization, hepatocellular necrosis, portal inflammation, fibrosis and cirrhosis.

[1] Supported by a Multipurpose Arthritis Center Grant AM-20621, an Arthritis Foundation Clinical Research Center Grant and a grant from the Arthritis Foundation, Connecticut Chapter.

Table I. Data on the 15 patients on long-term MTX at the start of therapy and at follow-ups I and II

	Initial	Follow-up I	Follow-up II
Mean (range) disease duration pre MTX, years	9.4 (1.5–32)	–	–
Mean (range) treatment duration, months	–	11 (2–25)	42 (20–67)
Mean (range) cumulative MTX dose, mg	–	591 (70–1,430)	2,021 (915–3,705)
Mean (range) weekly MTX dose, mg	11.1 (7.5–20)	15 (7.5–25)	13.3 (7.5–25)
Route, per os/i.m.	13/2	10/4[1]	8/7
Prednisone number of patients/dose, mg	5/8.9	–	7/6.7
Rheumatoid factor number of patients/titer	15/1:320–1:640	–	13/1:160–1:320

[1] One patient had dropped out but was subsequently restarted on MTX.

Results

Effectiveness

The clinical data on the 15 patients who were on MTX at this assessment is shown in table I. Of note is the absence of a statistically significant steroid-sparing effect or drop in rheumatoid factor titer. Assessment of disease activity is shown in table II. The clinical improvement noted at the first follow-up was continued through this assessment. Morning stiffness had completely disappeared in 8/15 (53%) patients and the sedimentation rate had normalized in 9/15 (60%) patients. Of the original 21 patients, global assessment (table III) revealed that 3 (14%) patients had achieved clinical remission [2]. Nine (43%) patients had an excellent response and 5 (24%) discontinued MTX because of side effects or fear of toxicity.

Toxicity

Adverse side effects were common (table IV). There were often temporary interruptions in therapy for gastrointestinal intolerance and abnormal liver function tests but permanent withdrawal of MTX was

Table II. Clinical variables at the start of MTX therapy and at follow-ups I and II in the 15 patients (mean ± SD)

	Initial	Follow-up I (11 months)	Follow-up II (42 months)
Tender joints, n	22.2 ± 10	13.1 ± 11*	5.9 ± 6**
Swollen joints, n	15.1 ± 8	6.7 ± 4*	3.9 ± 3**
Grip strength, mm Hg	82 ± 33	123 ± 49*	136 ± 51
AM stiffness, h	8.2 ± 5	3.4 ± 5*	0.4 ± 0.9
ESR, mm/h	68 ± 35	43 ± 22*	33 ± 25

* I vs. initial: $p \leq 0.006$; ** II vs. I: $p \leq 0.02$.

Table III. Global response (21 patients)

Remission	3 (14%)[1]
Excellent	9
Fair	4
Drop-out	5 (24%)

[1] One patient in remission had also withdrawn from therapy.

unusual. Patients often experienced disease flares if MTX was withheld for more than 3 weeks but responded to reinstitution of drug therapy. No patients in this study developed opportunistic infections, pulmonary hypersensitivity reactions or malignancy. Of the 3 patients who had dropped out prior to the first follow-up, 1 has since restarted therapy and has had an excellent response. Between follow-ups I and II, 4 or more patients discontinued MTX. Two patients had severe nausea, 1 had a planned pregnancy and 1 was in clinical remission and desired discontinuation of treatment.

Liver toxicity was assessed in 25 patients (the original 21 and the 4 other biopsied patients who had received greater than 1 g MTX). Abnormal transaminases occurred in 12/25 (48%) patients at some time during the course of treatment. These elevations usually first occurred 2–6 months after the start of MTX therapy. Seven patients had transaminase values greater than twice normal, sometimes up to 5 times normal. Recurrent milder elevations were seen in most (9/12) of these

Table IV. Adverse reactions (other than hepatotoxicity) and causes of permanent withdrawal from MTX in the 21 patients

Adverse reactions	n	Permanent withdrawal	n
None	9	Gastrointestinal	2
Gastrointestinal (nausea, dyspepsia, vomiting)	7	'Hypersensitivity' with hepatitis	1
		Noncompliance	1
Stomatitis	3	Planned pregnancy	1
Transient thrombocytopenia	1	Remission and fear of toxicity	1
Herpes zoster	1		
Acute hepatitis	1		

Table V. Results of liver histopathology in the 17 biopsied patients (cumulative MTX dose: mean 1,950 mg; range 915–3,125)

	Steatosis	Variation of nuclear size	Nuclear vacuolization	Hepatocellular necrosis	Periportal inflammation	Fibrosis
None	5	–	8	10	12	11
Minimal to mild	8	6	7	7	5	6
Moderate	4	10	1	–	–	–
Severe	–	1	1	–	–	–

patients throughout the study. The occurrence of abnormal transaminases was independent of the route of MTX administration.

Analysis of the results of the 17 liver biopsies is given in table V. While only 3 of the biopsies were completely normal, most revealed only minimal or mild pathological changes. Nonspecific changes with variation in nuclear size and nuclear vacuolization were prominent in some biopsies but were seen independently of more serious hepatic pathology and did not correlate with the occurrence of abnormal transaminase levels. Serious hepatic fibrosis or cirrhosis were not found.

Discussion

Recently, controlled trials have confirmed previous reports that low dose methotrexate is an effective short-term agent for the treatment of RA [1, 3–6]. This study extends our original observations and has shown that MTX has continued to be effective and safe after 42 months of therapy and a mean cumulative dose of over 2 g. Interestingly, 3 patients fulfilled ARA criteria for clinical remission and in 60% of the patients the sedimentation rate normalized. This suggests that MTX may prove to be a disease-modifying agent such as gold [7]. However, in most patients the benefit of MTX therapy receded quickly after drug withdrawal with clinical flares in 3–4 weeks. This suggests that MTX is inducing a suppression rather than a remission of the rheumatoid synovitis. Our results are in keeping with those of others [3, 4] which suggests that 50% of patients will experience long-term benefit from MTX.

MTX has been used intramuscularly [8] and intravenously [9] as well as orally, although there appears to be no therapeutic advantage to parenteral therapy. Although the optimum dose may vary from patient to patient, it is of interest that the mean dose at follow-up II was 13.3 mg weekly which was similar to the starting mean dose of 11.1 mg. Most of our patients are currently started on oral MTX at a dose of 10–15 mg as a single or divided dose once weekly.

Clinical side effects were common but mild. The absence of serious toxicity is likely a reflection of the small numbers of patients studied, the low doses of MTX being used and the relative safety of this regimen. While abnormal transaminases were seen in almost half the patients at some time during the course of treatment, the levels dropped with dose reduction and in most patients, levels normalized or remained only mildly and intermittently abnormal with continued therapy.

Previous reports have suggested that hepatic fibrosis and cirrhosis may occur in up to 25% of psoriatic patients on prolonged MTX treatment and that a cumulative dose of 1,500 mg is a significant risk factor [10, 11]. Despite a mean cumulative dose of 1,950 mg in the 17 biopsied patients, they showed no evidence of serious MTX toxicity. The changes that were seen are nonspecific and have been described in patients with RA who have not taken MTX [12]. Although serious liver damage may ensue with more prolonged therapy, the results of this study and 2 others reported in abstract [13, 14] suggest that MTX may be less hepatotoxic in RA than has been reported in psoriasis.

In conclusion, in a small number of RA patients followed for 42 months, MTX was effective at controlling synovitis with an acceptable incidence and severity of clinical side effects. Serious hepatic toxicity did not occur.

References

1 Steinsson, K.; Weinstein, A.; Korn, J.; Abeles, M.: Low dose methotrexate in rheumatoid arthritis. J. Rheumatol. *9:* 860–866 (1982).

2 Pinals, R.S.; Baum, J.; Bland, J.; Fosdick, W.M.; Kaplan, S.B.; Masi, A.T.; Mitchell, D.M.; Ropes, M.W.; Short, C.L.; Sigler, J.W.; Weinberger, H.J.: Preliminary criteria for clinical remission in rheumatoid arthritis. Bull. rheum. Dis. *32:* 45–48 (1982).

3 Willkens, R.F.; Watson, M.A.; Paxson, C.S.: Low dose pulse methotrexate therapy in rheumatoid arthritis. J. Rheumatol. *7:* 501–505 (1980).

4 Groff, G.D.; Shenberger, K.N.; Wilke, W.S.; Taylor, T.H.: Low dose oral methotrexate in rheumatoid arthritis: an uncontrolled trial and review of the literature. Semin. Arthrit. Rheumatism *12:* 333–347 (1983).

5 Ward, J.R.: Cooperative Studies of Rheumatic Diseases: A prospective randomized controlled trial of low dose pulse methotrexate (MTX) in rheumatoid arthritis (RA). Arthritis Rheum. *27:* suppl., p. 40 (1984).

6 Weinblatt, M.E.; Coblyn, J.S.; Fox, D.A.; Fraser, P.A.; Holdsworth, D.E.; Glass, D.N.; Trentham, D.E.: Efficacy of low-dose methotrexate in rheumatoid arthritis. New Engl. J. Med. *31:* 818–822 (1985).

7 Hunder, G.C.; Bunch, T.W.: Treatment of rheumatoid arthritis. Bull. rheum. Dis. *32:* 1–7 (1982).

8 Hoffmeister, R.T.: Methotrexate therapy in rheumatoid arthritis: 15 years experience. Am. J. Med., suppl., pp. 69–73 (Dec. 1983).

9 Michaels, R.M.; Nashel, D.J.; Leonard, A.; Sliwinski, A.; Derbes, S.J.: Weekly intravenous methotrexate in the treatment of rheumatoid arthritis. Arthritis Rheum. *25:* 339–341 (1982).

10 Nyfors, A.; Poulsen, H.: Liver biopsies from psoriatics related to methotrexate therapy. 2. Findings before and after methotrexate therapy. A blind study. Acta pathol. microbiol. scand., A, Pathol. *84:* 262–270 (1976).

11 Zachariae, H.; Kragballe, K.; Sogaard, H.: Methotrexate-induced liver cirrhosis. Studies including serial liver biopsies during continued treatment. Br. J. Derm. *102:* 407–412 (1980).

12 Lefkovits, A.M.; Farrow, I.J.: The liver in rheumatoid arthritis. Ann. rheum. Dis. *14:* 162–169 (1955).

13 Mackenzie, A.H.: Liver biopsy findings after methotrexate (MTX) therapy for rheumatoid arthritis (RA). J. Rheumatol. *1:* suppl., p.73 (1974).

14 DiBartolomeo, A.G.; Mayes, M.D.; Bathon, J.M.: Methotrexate (MTX) in rheumatoid arthritis (RA): a longitudinal study of liver biopsies. Arthritis Rheum. *27:* suppl., p. 61 (1984).

Prof. A. Weinstein, MD, Division of Rheumatic Diseases and Immunology, New York Medical College, Valhalla, NY 10595 (USA)

Rheumatology, vol. 9, pp. 154–169 (Karger, Basel 1986)

Methotrexate Therapy of Highly Active Polyarthritis

T. Karger, R. Rau

Department of Rheumatology, Evangelisches Fachkrankenhaus, Ratingen, FRG

After the first results of folic acid antagonist in the treatment of arthritis were reported in 1951 [10], it took many years until further results were published [5, 22].

Convincing experience of methotrexate (MTX) treatment in the nonmalignant disease psoriasis was established [4, 15, 28] after the low toxicity of the divided dose regimen had been proved [7]. MTX in low-dose application is considered a drug without carcinogenicity [3, 24, 28]. The possible risk in long-term treatment of nonmalignant disease remains unpredictable; some additional problems, especially regarding hepatotoxicity, are well known [16, 18, 34]. Important criteria for the therapeutic use of MTX were published in 1982 [19]. Until 1980, MTX was used rarely in the treatment of rheumatoid arthritis (RA) [11, 22]. The first uncontrolled trials revealed promising results in the treatment of intractable RA [30, 32]. In the last 2 years, numerous results have been reported [1, 2, 6, 8, 9, 12, 13, 20, 21, 23, 27, 35]; only one brief report showed poor long-term results [7].

Our own experiences with MTX date back to 1979. After having obtained convincing results in the treatment of psoriatic arthritis we started to also treat patients with rheumatoid arthritis in cases of very high disease activity. Because of the possible risk of a cytostatic therapy and considering the very rapid effect of the drug according to clinical and laboratory activity criteria, our philosophy during the first years was to introduce MTX only for a short period of time – 2–4 months – to overcome a phase of very high inflammatory activity of the disease. So MTX was prescribed in addition to a disease-modifying drug, for instance gold, which should be continued after MTX was stopped. In most cases MTX and gold were introduced while the patient was hospitalized. The family doctor was asked to continue MTX as a short-term and gold as a long-term treatment. After this more experimental phase and after becoming more and more familiar with the drug, we started to introduce MTX also as a long-term treatment in patients who did not respond at all to conventional forms of therapy.

Patients and Methods

Until March 1984 we introduced MTX treatment in 120 patients. 53 of these patients are not taken into consideration in this report because in most of these patients, as mentioned above, no long-term follow-up was planned and therefore a statistical evaluation was not possible.

25 of these 53 patients did not return to our outpatient department after commencing MTX therapy.

The remaining 28 patients are well documented in both hospital and outpatient records. In 7 of these 28 patients the reason for termination of MTX therapy was a complete remission of the disease. Side effects probably caused by MTX led to stopping treatment in 7 other cases. In the remaining 13 cases, MTX treatment was halted because the therapeutic goal of arresting an acute inflammatory phase had been achieved. In 1 patient treatment was stopped because of the development of a severe disease not corresponding with the RA or the MTX treatment.

In the following we report on 67 patients who received MTX as long-term therapy over a considerable period of time and who were given regular follow-up examinations. Our 67 patients can be divided into two groups: The first group of 31 patients (group A) was first re-examined at periods varying from 1 to 24 months after the start of treatment. Prospective follow-ups were then carried out at 3-month intervals. Group B included 36 patients who received prospective examinations 1, 2, 4 and 6 months after treatment began and then every 3 months.

The results of 6 patients were ignored in the statistical evaluation because of non-compliance during the course of therapy.

The treatment was usually started with a weekly dose of 25 mg administered i.v. for 3 or 4 weeks and then continued with a weekly dose of between 5 and 15 mg (depending on the clinical and subjective findings) taken orally as a divided dose in three portions at 12-hour intervals. A constant dose of NSAID was allowed. The dosage of steroids, whether in a systemic or an intra-articular mode of application was determined individually. 50% of all patients were initially treated with parenteral gold salts; in most cases gold treatment was discontinued during the course of MTX treatment.

In each of the re-examinations the following results were recorded: subjective judgement of pain and motility, grip strength and joint count (synovitis was given 1 point, synovitis and joint effusion 2 points).

The examiner rated the patient according to Voit-Gamp's activity index (table I) and ARA functional class. We did not include an assessment of morning stiffness, since it seemed too imprecise a factor for measurement purposes.

Table I. Activity index according to *Voit and Gamp* [26]

Index of activity	Deterioration of the general condition (loss of weight, fever)	Development of rheumatoid disease	Inflammation of joints	ESR	X-ray
1 No activity	0	no involvement of new joints	no synovitis	< 10–20	no progression of radiological signs
2 Low activity	0	no new extra-articular manifestations, reduced motility possible	low inflammatory joint symptoms	20–60	progression of radiological signs possible
3 Moderate activity	low	progression possible	moderate joint symptoms in few joints	20–60	progression of radiological signs possible
4 Severe activity	severe	other joints affected, new extra-articular manifestations	severe synovitis in many joints	> 60	progression of radiological signs

The following laboratory results were recorded regularly: ESR, differential blood count, Quick value, GOT, GPT, γ-GT, AP, creatinine and urinalysis for blood and protein. Every half year the following data were measured: rheumatoid factor (using the latex test), Rose-Waaler, total serum protein, electrophoresis of serum proteins, acid-1-glycoprotein and immunoglobulins.

The average age of the patients was 53.5 years (39–73). 18% were male and 82% female. The average disease duration preceding therapy was 8.36 (± 6.6) years, minimum 1 and maximum 34 years.

Diagnoses

59 patients (88%) suffered from rheumatoid arthritis, 49 patients (83%) of these were seropositive. Four patients (6%) suffered from psoriatic arthritis, 1 patient from Reiter's disease, 1 patient from adult onset Still's disease, and 2 patients from overlap syndrome. 38% of the RA patients had extra-articular manifestations of the disease (vasculitis, pericardial effusion, rheumatoid nodes). 49% of all patients had had one or more radiosynovectomies. 46% had been operated on one or more joints; 10% had had more than 5 joint operations. Anatomical stages and functional classes of the patients are listed in table II.

With the exception of 2 patients who were suffering from a sudden onslaught of psoriatic arthritis, all patients had been treated with gold salts. In about half of these

Table II. Anatomical stage and functional class of all 67 patients (group A and group B) before MTX treatment was started

ARA anatomical stage				ARA functional class			
I	II	III	IV	1	2	3	4
0%	31%	47%	22%	2%	63%	31%	5%

Table III. Therapy before MTX treatment was started

	Not yet, %	Simultaneous with MTX, %	Discontinued because of	
			side effects %	lack of response, %
Gold	3	51	23	23
d-Penicillamine	66	2	19	14
Chloroquine	75	–	9	16
Others[1]	66	–	11	23

[1] Cyclophosphamide, ifosfamide, chlorambucil, mitopodocide, azathioprine.

cases treatment had to be stopped because of side effects or lack of improvement. 34% had been previously treated with *d*-penicillamine, 25% with chloroquine and 30% with immunosuppressive drugs, either unsuccessfully or with side effects (table III).

Statistical Evaluation

The statistical evaluation was performed by Prof. *Neiss,* Gesellschaft für Mathematik und Informatik, München.

The Wilcoxson test, the sign test and the McNemar test were used. For group A the data after 2 years and again after 3 years of treatment were statistically compared with those recorded at the beginning of therapy. For group B statistical evaluation was made after 1, 2, 6 and 12 months of treatment.

Results

Whereas at the start of therapy no patient rated *pain and motility* as good, after only 1 month of therapy was a significant improvement seen in patients' assessment (fig. 1). This improvement remains con-

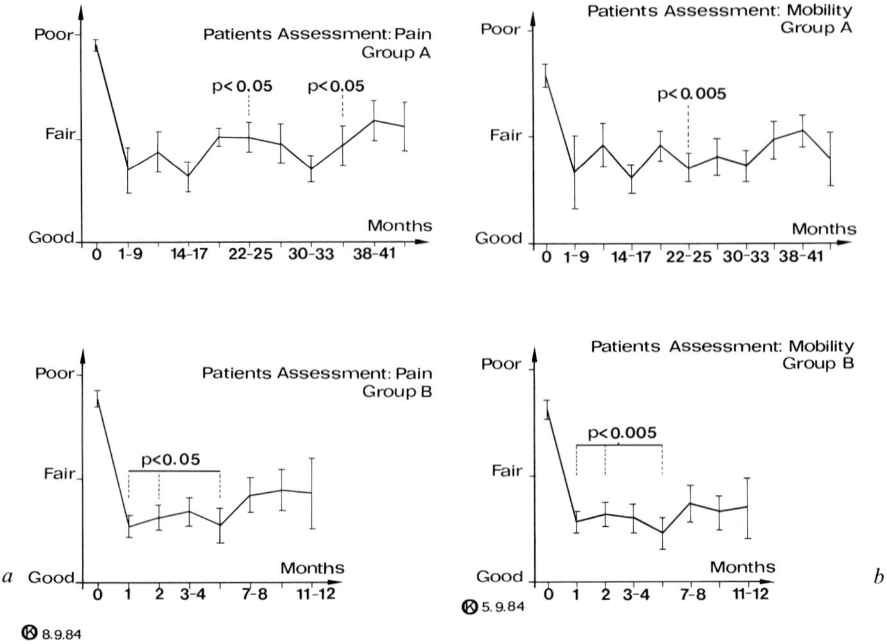

Fig. 1. Improvement of pain (*a*) and mobility (*b*) in the patients assessment. Significant changes are marked. After 3 years (group A) and after 6 months (group B), the changes reached no significance because the number of patients was too small.

stant over the rather long period of time and is reflected in the examiners equally subjective assessment giving a *Steinbrocker functional class* (fig. 2). The degree of *disease activity* improves also significantly in both groups (fig. 3). Of particular importance is the fact that the patients' positive assessment of their condition continues throughout the period of observation as well as the activity index and the functional class.

The *joint count* (fig. 4) is significantly lowered and here too the improvement continues. *Grip strength* of both hands is significantly increased (fig. 5).

An important indicator of effectiveness is the *reduced need for steroid medication:* in both groups the average daily needed dose of prednisone could be reduced significantly below 4.1 or 5 mg, respectively, which had been administered previously (fig. 6).

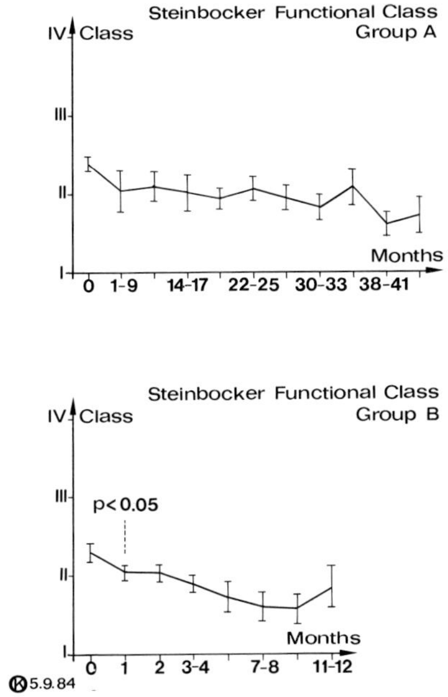

2

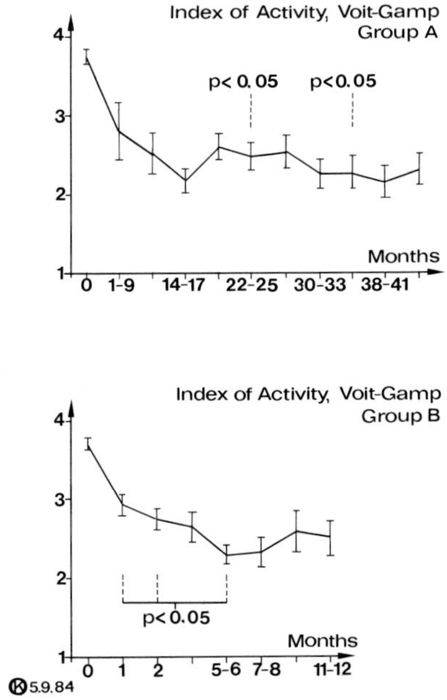

3

Fig. 2. Changes of the Steinbrocker functional class.

Fig. 3. Changes of disease activity index according to Voit-Gamp.

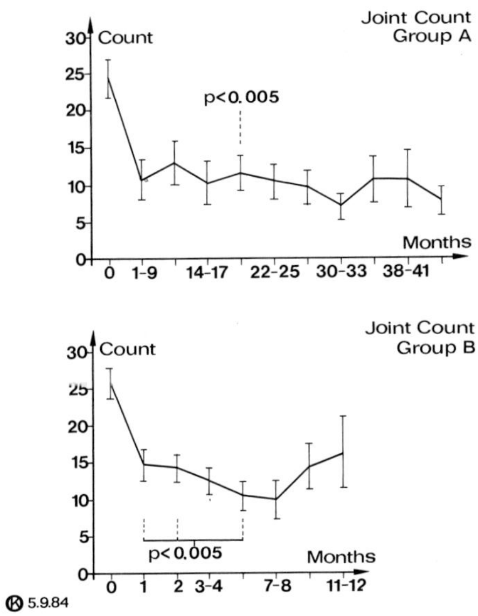

Fig. 4. Changes of the joint count. All affected small and large joints were counted. Synovitis was counted as one point and synovitis with effusion as two points.

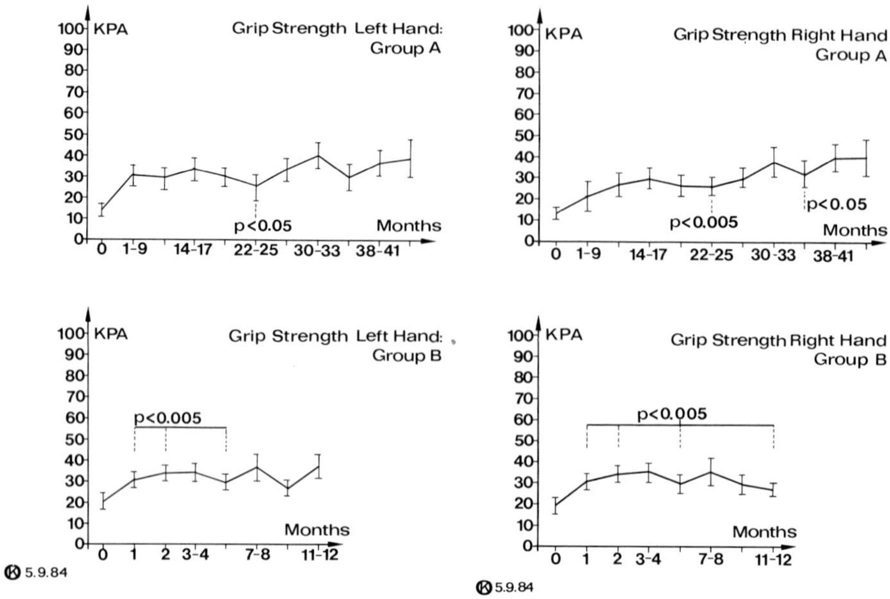

a b

Fig. 5. Changes of the grip strengh of both hands. The values are given in kP/cm². The test was performed with a Vigorimeter® (Martin).

The reduction in the total number of intra-articular steroid injections given each year was even more impressive (fig. 7).

The *ESR* was significantly influenced in both groups (fig. 8). No significant changes were observed in any other laboratory tests, although there was a continual fall in the platelet count from increased level to normal and a slight increase in the hemoglobin values as well as a drop in acute phase reactants. We did not see a measurable effect on protein electrophoresis and on the concentration of immunoglobulins, possibly due to the wide tolerance of these methods. There were no signs of myelosuppression, nor of renal damage or cystitis.

In 23 cases therapy was withdrawn (table IV). In only 2 of these cases were untoward effects attributable to MTX the cause for discontinuation of treatment. In all other cases other factors were responsible for the cessation of therapy.

Side effects were noted by the patients' complaints as spontaneous complaints of the patients and in addition by a special inquiry (table V). The side effects listed in table V were reported in a total of 385 reexaminations. The most frequent adverse effects (group A + group

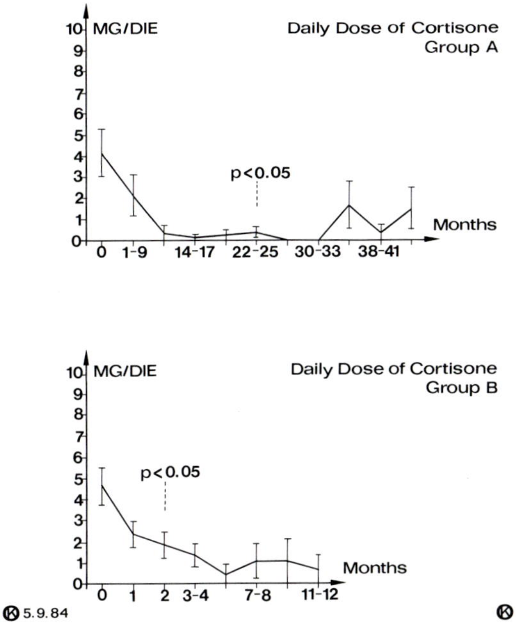

6

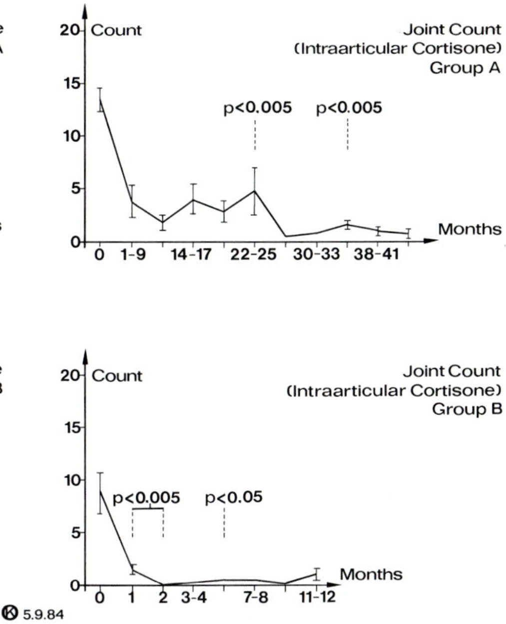

7

Fig. 6. Changes of the daily intake of prednisone.

Fig. 7. Changes of the number of intra-articular injections of corticosteroids (triamcinolone-acetonide in most cases) given per year. The number of i.a. injections before starting MTX treatment was recorded in our outpatient clinic and calculated for 1 year.

B) were nausea (44 patients/93 times), GI distress (22 patients/24 times) and loss of hair (23 patients/30 times). Less common side effects were stomatitis, skin lesions, pulmonary symptoms, increased pain after taking MTX, as well as blurred vision, burning eyes and dizziness.

All side effects usually appeared in a mild form and were tolerated by the patients. It was possible to continue therapy after reducing the dosage or interrupting treatment for a short time. We did not see more or different complications in simultaneous application of MTX and gold salts: 23 patients were treated simultaneously up to periods of 40 months: 1 patient had a severe dermatitis after 20 months, 1 patient complained of ageusia after 12 months and 2 patients had a stomatitis after 4 months of therapy.

Certain uncommon side effects were recorded (table VI) which in the light of the patient's personality and the surrounding circum-

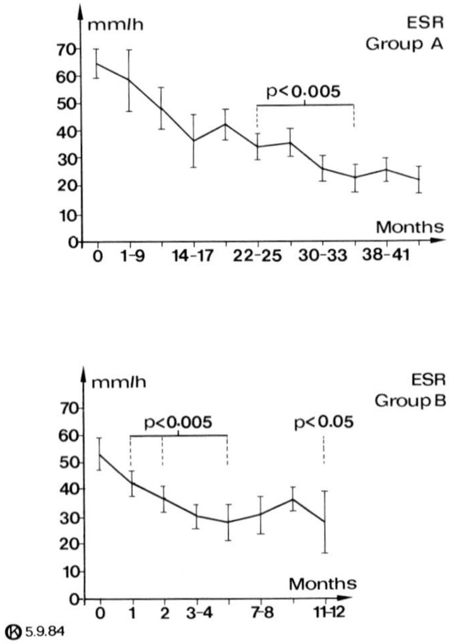

Fig. 8. Changes of the ESR (1 h, Westergren method).

stances cannot necessarily be attributed to MTX. The labelling of clinical symptoms as side effect is a subjective judgement on the part of the examiner, and it is not always possible to decide whether the adverse effects are a result of MTX therapy, of the concomitant therapy or of the underlying illness.

We often observed a slight to moderate increase in *liver enzymes:* in a total of 32 patients increased GPT occurred in 51 examinations up to 4 times normal. In 3 patients we found an elevated GPT up to 300 IU/l. In 13 patients 28 examinations revealed elevated γ-GT with a maximum of 100 IU/l. In 18 patients elevated AP was observed in 23 examinations; in 9 tests by more than 300 IU/l. In all cases the measured values were reversible after a short break in treatment or a dose reduction.

The case history (fig. 9) indicates that even when therapy is continued there is a spontaneous decrease in liver enzymes.

Some examples of good clinical and serological response to MTX treatment are shown in figure 10.

Table IV. Reasons for discontinuation of MTX therapy in 23 patients

Number of patients	Reason	Number of patients	Reason
3	Remission	6	On patient's decision (apprehension of the drug)
5	Drop out	2	Side effects
4	No compliance	3	Other reasons

Table V. Frequency of side effects recorded in a total of 385 re-examinations

	Group A (n = 29)		Group B (n = 33)	
	number of patients	frequency of complaints	number of patients	frequency of complaints
Nausea, vomitus	19 (65%)	45	25 (69%)	48
Diarrhea, abdominal pain	10 (34%)	11	11 (31%)	13
Loss of hair	14 (48%)	20	9 (27%)	10
Stomatitis	6 (21%)	6	6 (18%)	6
Skin lesions	2 (7%)	2	6 (18%)	6
Pulmonary symptoms	2 (7%)	2	2 (6%)	2
Increased pain	1 (3%)	1	1 (3%)	1
Others	9 (31%)	10	8 (24%)	11

Table VI. Unusual side effects, not unequivocal due to MTX

Symptoms	Special features
2 × increased pain after MTX intake	noncompliance in both patients, 1 had IgA deficiency
1 × duodenal ulcer	after 115 mg MTX; patient had history of GI ulcer; 3 weeks before MTX was started she had been operated for a breast cancer
1 × daze, impaired vision, apathy	noncompliance, apprehensive of all therapy
1 × pyodermatosis, dose dependent	patient with psoriasis; IgA deficiency, severe dermatitis after gold treatment
2 × GI distress in oral therapy of MTX; no symptoms in i.v. therapy	

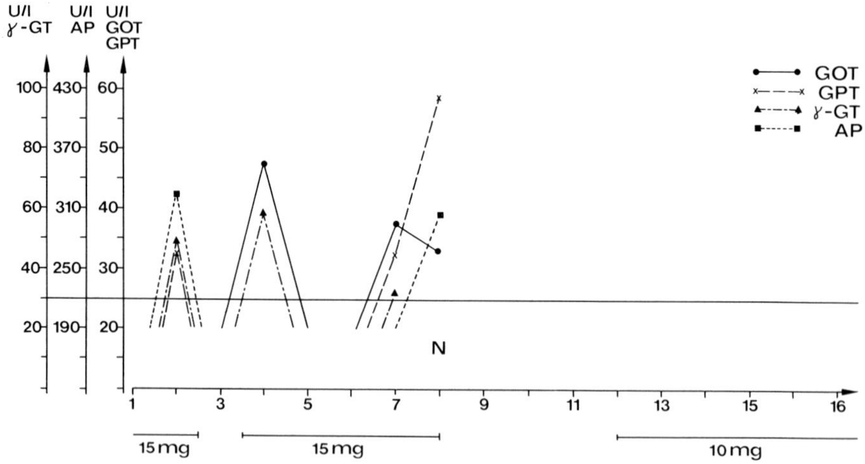

a

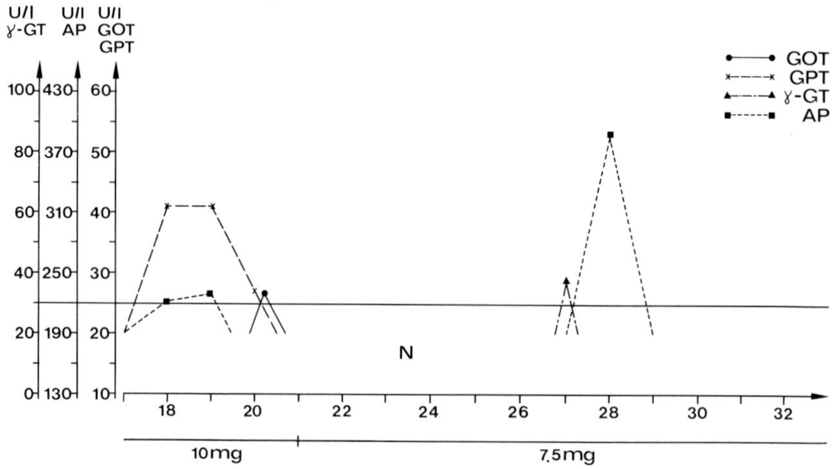

b

Fig. 9. Case history: elevation of liver enzymes during MTX treatment in a patient with rheumatoid arthritis.

Fig. 10. Changes of ESR and joint count in 6 patients after starting MTX therapy. In 3 patients, who had prednisone, the corticosteroid application could be stopped.

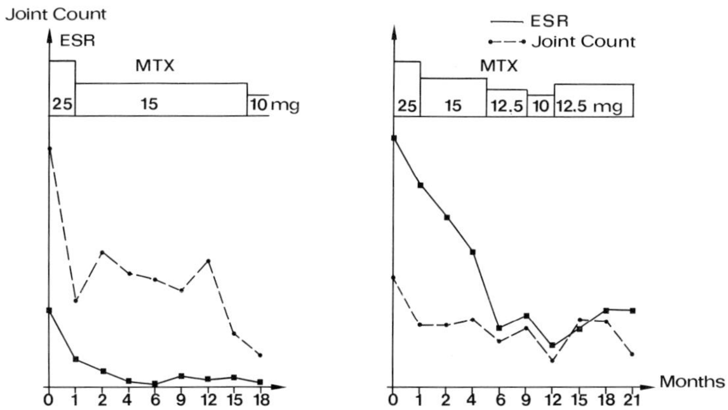
Joint Count
ESR
MTX
25
15
10 mg
0 1 2 4 6 9 12 15 18
ESR
Joint Count
MTX
25
15
12.5
10
12.5 mg
0 1 2 4 6 9 12 15 18 21
Months

a

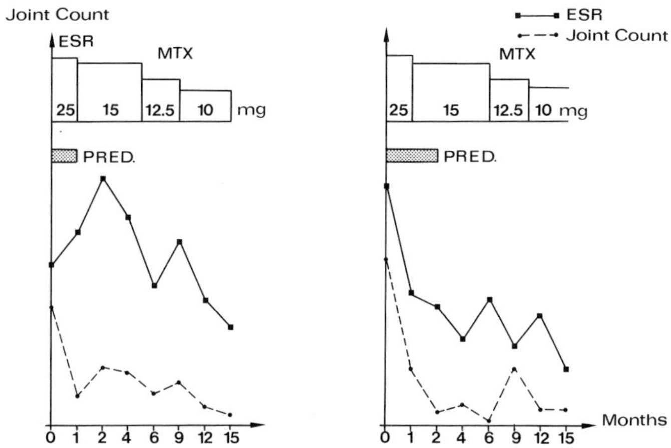
Joint Count
ESR
MTX
25
15
12.5
10
mg
PRED.
0 1 2 4 6 9 12 15
ESR
Joint Count
MTX
25
15
12.5
10 mg
PRED.
0 1 2 4 6 9 12 15
Months

b

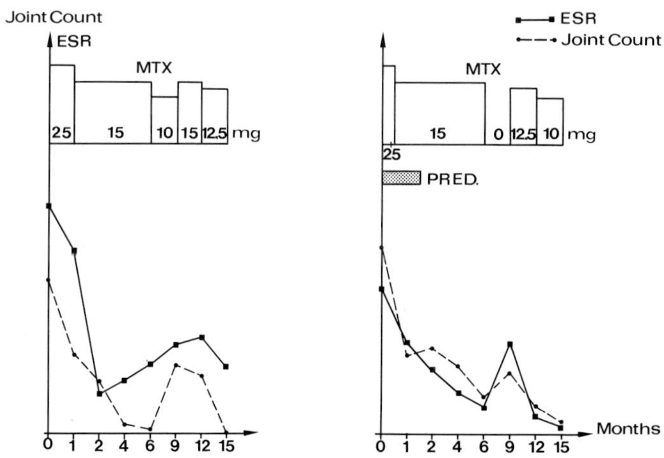
Joint Count
ESR
MTX
25
15
10
15
12.5
mg
0 1 2 4 6 9 12 15
ESR
Joint Count
MTX
15
0
12.5
10
mg
25
PRED.
0 1 2 4 6 9 12 15
Months

c

Table VII. Cumulative dose and mean needed weekly dose of MTX in the long-term treatment (mg values)

	Cumulative dose	Per week
Group A		
12 months	632 (410–1,000)	12.15
24 months	1,040 (345–1,475)	7.8
36 months	1,460 (560–1,895)	8.0
Group B		
6 months	411 (250–660)	15.8
12 months	750 (590–1,060)	13.04

In 16 cases with signs of remission or where NSAIDs were no longer required, we suspended MTX therapy. Usually, after 4–6 weeks in all cases disease activity flared up again and MTX was reintroduced. The longest break in therapy was 5 months.

In a few cases MTX treatment with a weekly dose of between 5 and 12.5 mg is the only form of therapy used. NSAIDs are either seldom or never necessary. As a result of individual dosage and the possibility of MTX-free intervals, there is a continual reduction in the needed mean weekly dose in long-term treatment (table VII).

Conclusions

The results of our study indicate that low-dose MTX exerts a very rapid and pronounced effect on the inflammatory signs and symptoms of rheumatoid arthritis. Already within 2 months of treatment joint count, grip strength, daily needed steroid dose and ESR, as well as the subjective rates of patient and physician, were significantly improved and this improvement could be maintained throughout the duration of treatment. The fact that within a few weeks after discontinuation of MTX therapy a relapse of symptoms could be observed in most cases seems to indicate that the mode of action is more anti-inflammatory than disease modifying.

During the first year of treatment the mean weekly doses were between 10 and 15 mg, during the second and third year between 5 and 10 mg of MTX.

MTX seems to be a relatively save drug, compared with other cytotoxic compounds and other disease-modifying forms of treatment, respectively. Side effects are in general mild, reversible and dose related [31], most commonly effecting the gastrointestinal tract. Elevated liver enzymes, which can be observed in active rheumatoid arthritis also without treatment as an indicator of a nonspecific reactive hepatitis [17], occur in almost 50% of patients with MTX therapy but are reversible even without interruption of treatment and, as far as we have seen, do not cause structural changes of the liver. Bone marrow suppression, renal disease or malignancies, respectively, have not been observed. No serious outcome occurred in our patients.

An absolutely necessary prerequisite of this treatment is that the patient is given extensive information and strictly follows the physicians plan of treatment, in other words: the commitment of the doctor must win the commitment and trust of the patient.

We agree with a statement [25] that MTX is a promising drug in the treatment of rheumatoid arthritis and other polyarthritic diseases. Further double-blind trials [33] and long-term follow-up investigations together with an evaluation of the radiological changes are needed to prove the effectiveness and tolerance of the drug.

References

1 Andersen, P.A.; West, S.G.; Clapool, R.C.: Pulse methotrexate therapy for rheumatoid arthritis. A double-blind, crossover study (Abstract). Arthritis Rheum. *27:* suppl., p. S25 (1984).

2 Bachmann, D.M.: Pulsed intravenous methotrexate treatment in rheumatoid arthritis (Abstract). Arthritis Rheum. *25:* suppl., p. S65 (1982).

3 Bailin, P.L.; Tindall, J.P.; Roenigk, H.H.; Hogan, M.D.: Is methotrexate for psoriasis carcinogenic? J. Am. med. Ass. *232:* 359–362 (1975).

4 Bergstresser, P.R.; Schreiber, S.H.; Weinstein, G.D.: Systemic chemotherapy for psoriasis. Archs. Derm. *112:* 977–981 (1976).

5 Black, R.L.; O'Brien, W.M.; Scott, E.J.: Methotrexate therapy in psoriasis arthritis. J. Am. med. Ass. *189:* 743–744 (1964).

6 DiBartolomeo, A.G.; Mayes, M.D.; Bathon, J.M.: Methotrexate in rheumatoid arthritis. A longitudinal study of liver biopsies (Abstract). Arthritis Rheum. *27:* suppl., p. S61 (1984).

7 Fitzgerald, O.; Hanly, J.; Molony, J.; Bresnihan, B.: Poor long-term results from low-dose methotrexate therapy in rheumatoid arthritis. Arthritis Rheum. *27:* 599–600 (1984).

8 Gowans, J.D.C.; Kaplan, A.G.; Chang, W.G.: Long-term therapy of psoriatic and rheumatoid arthritis with oral methotrexate monitored by serial liver biopsies (Abstract). Arthritis Rheum. *22:* 615–616 (1979).

9 Groff, G.D.; Shenberger, K.N.; Wilke, W.S.; Taylot, T.H.: Low dose oral methotrexate in rheumatoid arthritis: an uncontrolled trial and review of the literature. Semin. Arthrit. Rheumatism *12:* 333–347 (1983).

10 Gubner, R.S.: Effect of aminopterin in rheumatoid arthritis and psoriasis. Am. J. med. Sci. *221:* 176 (1951).

11 Hoffmeister, R.T.: Methotrexate in rheumatoid arthritis. Arthritis Rheum. *15:* 114 (1972).

12 Karger, T.; Rau, R.: Methotrexat-Behandlung hochaktiver chronischer Polyarthritiden. Münch. med. Wschr. *126:* 560–564 (1984).

13 McKendry, R.; Hamdy, H.; Mierins, E.: Azathioprine compared with methotrexate in the treatment of rheumatoid arthritis. Results of a double-blind, 24 week trial (Abstract).Arthritis Rheum. *27:* suppl., p. S57 (1984).

14 Michaels, R.M.; Nashel, D.J.; Leonhard, A.; Sliwinski, A.J.: Weekly intravenous methotrexate in the treatment of rheumatoid arthritis. Arthritis Rheum. *25:* 339–341 (1982).

15 Nyfors, A.: Methotrexate therapy of psoriasis. Dan. med. Bull. *27:* 74–96 (1980).

16 Nyfors, A.; Poulsen, H.: Liver biopsies from psoriatics related to methotrexate therapy; in Farber, Cox, Psoriasis. Proc. 2nd Int. Symp., Stanford University 1976, pp. 235–243.

17 Rau, R.: Die Leber bei entzündlichen rheumatischen Erkrankungen (Steinkopff, Darmstadt 1978).

18 Robinson, J.K.; Baughman, R.D.; Auerbach, R.; Cimis, R.J.: Methotrexate hepatotoxicity in psoriasis. Archs. Derm. *116:* 413–415 (1980).

19 Roenigk, H.H.; Auerbach, R.; Maibach, H.J.; Weinstein, G.D.: Methotrexate guidelines – revised. J. Am. Acad. Derm. *6:* 145–155 (1982).

20 Roth, S.H.: Comparison of pulse methotrexate therapy with gold salt therapy in rheumatoid arthritis (Abstract). Arthritis Rheum. *24:* suppl., p. S71 (1981).

21 Russell, A.S.; Watts, C.; Thompson, R.; Edelman, J.; Esdaile, J.; Percy, J.; Davis, P.; Tannenbaum, H.; Grace, M.: A double blind placebo-controlled short-term trial of parenteral methotrexate in rheumatoid arthritis (Abstract). Arthritis Rheum. *27:* suppl., p. S59 (1984).

22 Siegmeth, W.; Eberl, R.; Tausch, G.; Raith, W.: Vergleichende Untersuchung der Wirkung des Antimetabolicums Methotrexat bei Psoriasis, Arthropathia psoriatika und rheumatoider Arthritis. Verh. dt. Ges. Rheumatol. *1:* 384–389 (1969).

23 Steinsson, K.; Weinstein, A.; Korn, J.; Abeles, M.: Low dose methotrexate in rheumatoid arthritis. J. Rheumatol. *9:* 860–866 (1982).

24 Turnbull, C.; Roach, M.: Is methotrexate carcinogenic? Br. med. J. *281:* 808 (1980).

25 Twenty-fifth rheumatism review. Arthritis Rheum. *26:* 300 (1983).

26 Voit, K.; Gamp, A.: Der Rheumatismus (Stuttgart 1958).

27 Ward, J.R.: A prospective randomized controlled trial of low dose pulse methotrexate in rheumatoid arthritis (Abstract). Arthritis Rheum. *27:* suppl., p. S40 (1984).

28 Weinstein, G.D.: Methotrexate. Ann. intern. Med. *86:* 199–204 (1977).

29 Weinstein, G.D.; Frost, P.: Methotrexate for psoriasis. Archs. Derm. *103:* 33–38 (1971).
30 Wilke, W.S.; Calabrese, L.H.; Scherbel, A.L.: Methotrexate in the treatment of rheumatoid arthritis. Cleveland Clin. Q. *47:* 305–309 (1980).
31 Wilke, W.S.; Mackenzie, A.H.; Scherbel, A.L.; Groff, G.D.; Taylor, T.H.: Toxicity from methotrexate may be dose related. Arthritis Rheum. *26:* 119 (1983).
32 Willkens, R.F.; Watson, M.A.; Paxson, C.S.: Low dose pulse methotrexate therapy in rheumatoid arthritis. J. Rheumatol. *7:* 501–505 (1980).
33 Willkens, R.F.; Williams, H.J.; Ward, J.R.; Egger, M.J.; Reading, J.C.; Clements, P.J.; Cathcart, E.S.; Samuelson, C.Q.; Solsky, M.A.; Kaplan, S.B.; Guttadauria, M.; Halla, J.T.; Weinstein, A.: Randomized double-blind, placebo controlled trial of low-dose pulse methotrexate in psoriatic arthritis. Arthritis Rheum. *27:* 376–381 (1984).
34 Zachariae, H.; Kragballe, K.; Søgaard, H.: Methotrexate induced liver cirrhosis. Br. J. Derm. *102:* 407–412 (1980).
35 Zeiders, R.S.: Oral methotrexate therapy in rheumatoid arthritis (Abstract). Arthritis Rheum. *25:* suppl., p. S65 (1982).

Dr. T. Karger, Department of Rheumatology, Evangelisches Fachkrankenhaus, Rosenstrasse 2, D-4030 Ratingen (FRG)

Rheumatology, vol. 9, pp. 170–173 (Karger, Basel 1986)

Discussion 7

Willkens: I might ask Dr. *Weinstein* an initiating question: Do you mean to say that you were disregarding elevations of the transaminases and, if so, at what level would you disregard them or not disregard them?

Weinstein: Well, I think it's premature for me to say that I would disregard them. But most of the patients who had very significant elevations had it at around 3 months of therapy, 2–4 months, in that range. And we did make adjustments in dose, and either switched to i.m. or to lowering the dose or discontinuing it temporarily.

I disregard transaminase elevations in that I think they are irrelevant to the serious hepatic pathology that we are worried about with methotrexate.

In fact, now we are going up to three times normal. Over three times normal we are still making adjustments in dosage, or withdrawing temporarily or switching to i.m. But that's empirical.

Rau: Do you have any observations on the radiological development of the disease during these 41 months of treatment? That means, the development of erosions, for instance?

Weinstein: I have X-rays of all follow-up periods of all these patients. This is a very difficult subject to even interpret. I also have a cohort of patients not on methotrexate. This is going to be a subject of a separate study. I don't have any data to give you. Some patients got worse, some better, some erosions healed, at the same time some were getting worse. It's a very difficult thing to examine, I'm not sure if I get anything meaningful out of it, radiologically. But I have the X-rays.

Willkens: I would like to ask both Dr. *Rau* and Dr. *Karger,* whether you think, and I'm asking you this well in view with your experience with liver biopsies and liver assessment in patients with rheumatoid arthritis, do you think that nuclear vacuolisation is distinguishable from any effect that you think might be induced by methotrexate in direction to fibrosis?

Rau: I would like to pass the question to Dr. *Frenzel,* who is a pathologist.

Frenzel: I think, the vacuolation of the liver cell nucleus is a nonspecific phenomenon. We see these alterations in a lot of situations, for instance, in diabetics and in all the young patients. Patients at the age of 20 very often have vacuolated nuclei, and in Wilson's disease, too. Therefore, the vacuolation cannot be correlated with he toxic effect of methotrexate.

Willkens: But that's what I'm saying, it's distinguishable, it's very distinguishable from fibrosis and certainly the changes, the inflammatory cell infiltrations that we are now associating with methotrexate. The vacuolation is something quite different.

Frenzel: Yes, I think so.

Rau: One question to Dr. *Willkens.* If you stop treatment with MTX, you always, and Dr. *Michaels* says the same, and Dr. *Mackenzie* had the same impression, Dr. *Weinstein* as well, if you stop giving MTX, then you have a relapse of the rheumatoid arthritis within a short time. Would that mean that the effect is only symptomatic, or what do you think?

Willkens: I thought that it was not a remission-inducing drug, whether or not that defines an agent as being anti-inflammatory, versus an immunosuppressive, versus a remittive, I don't know. But certainly, all I can say is, you got to keep giving to sustain the effect.

Another question for you, are you proposing, with your dosage regimen, that we really need a loading dose of methotrexate prior to the institution of a maintenance dose, is that what you are proposing with your initial parenteral high dose and retreating to a low dose?

Rau: I think, you will get a much faster effect if you give 25 mg intravenously, and those patients are hospitalized in most cases, and there we can have a very, very fast effect. And after 2, 3 or 4 weeks they go to the out-patient clinic and they have a remission already, a remission in their disease, you see. You saw the graphs of Dr. *Karger,* within 1–2 months you have a remission already, and this effect can be maintained by the oral dose.

Dunky: I would like to join in this question. You give a higher starting dose, then you go back to the lower dose. At what time do you see the most side effects. At the beginning or later?

Karger: I think they aren't time dependent... we couldn't observe any time- or dosage-dependent frequency of side effects. There are patients who complain about nausea, or vomiting after 12 months as well as after 4 weeks.

Schmidt: Is there any different indication for giving lower doses or higher doses? Is there a different mechanism of action? Did you think that the initial high doses may also have an antiproliferative effect on the synovial membrane or is it also only an immunosuppressive or antiphlogistic effect? Because we all know that what we need is not a new antiphlogistic drug but an agent which is capable of inhibiting the proliferation of the synovial membrane. Is there any evidence that MTX could be able to act in such a manner?

Karger: As far as we know, there is no knowledge about that. There are some investigations about immunosuppressive mechanisms, we discussed that yesterday, and about anti-inflammatory mechanisms, but I don't know any publication about antiproliferative mode of action.

Willkens: Dr. *Furst* is actually doing, as I believe, a biopsy study, a synovial biopsy study in conjunction with his dosage of methotrexate, and perhaps this afternoon he'll come back to that.

Weinstein: Can I just make a comment about your comment on remittive properties. Wouldn't you say that no drug that we use, including gold, has remittive properties by your criteria? That is, if we stop the drug the patient stays in remission.

Willkens: Well, on the other hand, it's possible to induce some sustained, if you don't want to use the word remission, a more sustained decrease in symptoms, without giving gold for a period of time. I mean, at least the first time you give gold, you frequently come to point where you can stop it for a prolonged period of time and the disease will not be apparent.

Weinstein: Yes, it takes 6 months to have an effect and another 6 months for the disease to become active again.

Willkens: O.K., well, sure, I don't think we have a remission-inducing agent. I think it's a misuse of the term.

Dunky: There is also a question to Dr. *Karger.* You mentioned at last something about the compliance of the physicians. And you changed, that's my impression, you changed very often your dosing. 25 mg in the beginning and then you go back to 15, 12.5 and 10 mg. And I don't see how you can explain this to the family doctor, when you change and how you change.

Karger: The weekly dose is adapted by clinical results. If the joint count is dropping, if the ESR is dropping, and the patient is feeling well, you can make a dose reduction and every doctor, every family doctor will get information after a visit in the out-patient department. Written information or sometimes by phone.

Schmidt: As I understand, you gave MTX together with gold in about 50% of your patients. Were there more side effects in this group of patients?

Karger: No, I have a list here. 9 patients received from 20 to more than 40 months both gold and MTX. And of these 9 patients, 1 had a severe dermatitis after 20 months, and both gold and MTX were withdrawn. 6 patients had both drugs for 10–20 months, and 1 patient complained of dermatitis; 8 patients had gold and MTX together for up to 10 months. There were 2 patients with stomatitis: in 1 patient MTX and gold were withdrawn, in 1 patient only gold was stopped. MTX was continued and she is feeling well after 15 months of only MTX treatment.

Rau: Maybe this will be a way for future treatment of rheumatoid arthritis. As you know, we suppose that both drugs have a different way or a different mechanism of action, and so we can hope, as Dr. *Wilke* says, that we can potentiate the effect of both drugs, and perhaps reduce the dose of both drugs and this way get less side effects. You know the combination therapy is used by oncologists, but in rheumatology it's a rule not to combine two so-called remittive agents. But I don't know if this shall be a principle for eternity.

Willkens: One moment. I would like to get back just for a moment. We said that it's terribly important that we think about SGOT and maybe SGPT and liver toxicity. In *Weinstein's* papers, 20–30% of patients who had abnormal biopsies, which might include fibrosis and cirrhosis, didn't have elevated enzymes. So that means that you can miss it. Even though the enzymes are perfectly normal.

One other point. Some people, *Weinstein,* seem to think that 1.5 g is kind of a magic number. Above that, you see an increased incidence of hepatic toxicity. *Roenigh* was unable to show that that was true in another fairly large study, but at least there is a number there. I would be much more tempted to biopsy the liver of a patient who had a higher cumulative dose of the drug, and now has enzyme elevation. I would be much more interested in biopsying the patient who had a higher cumulative dose. Well, we are going to hear this question discussed again this afternoon. Dr. *Nyfors* and Dr. *Mackenzie* will give us some more data, maybe we can delay the discussion until then.

Mitrenga: Dr. *Karger,* I would like to ask you, in your group of highly active patients with rheumatoid arthritis, was there lung or cardiac involvement and what did methotrexate do in these cases?

Karger: Yes, if you remember the second or third slide, there was an extra-articular involvement in about 50% of the patients, and there was cardiac effusion in 2 or 3 patients.

Weinstein: I have a question for Dr. *Karger* and Dr. *Rau*. Because of the suggestion that methotrexate may not affect nodule formation, it may even worsen it, also it might suggest that in fact it has no effect on some of the extra-articular manifestations, including vasculitis. Now there have been, I noticed various slides suggesting that some of these patients had mononeuritis multiplex and it helped. I wonder if you have any organized data on the effect that methotrexate has on some of the extra-articular features of rheumatoid disease. Especially rheumatoid vasculitis.

Rau: I think we have no exact data on this topic.

Karger: No, there are no patients with typical rheumatoid vasculitis. One very impressive patient was a patient with cardiac effusion, she had a severe flare up with fever and cardiac effusion and was put on prednisone, about 30 mg/day, and she could be withdrawn from prednisone after 5 weeks and is now maintained on 7.5 mg of MTX for nearly 20 months, and is feeling completely well. No effusion, no other symptoms.

Wilke: I have got 4 dirty cases that come to mind immediately. Why are they dirty?

They showed up with rheumatoid vasculitis, all biopsy proven, all of them with mononeuritis multiplex, all with rheumatoid arthritis. They were first treated with nitrogen mustard in the same way that other people might have used intravenous cyclophosphamide. Using a dose of something like 3/10 of mg/kg delivered over around 10 days. So we spare the neutrophils. And then approximately 3 weeks later, all of these patients were put on methotrexate. Usually at about the biggest dose we give, which is around 12.5 mg. All 4 of these patients, who, up until that point, needed to be taking doses of prednisone in the range of 60 mg/day, have been followed for various periods of time, 1 was only followed for the past 4 months. One man died of a myocardial infarction 6 weeks ago. They all maintained very well on the methotrexate. We were able to lower the dose of prednisone. Of course, it's a dirty case, we've used an alkylating agent for a certain period of time to really get to the problem. We have not had to put them back on any alkylating agent after that first blast of nitrogen mustard.

So, we had seen vasculitis improve with methotrexate. In fact, we wonder whether methotrexate doesn't help us in steroid spare and treat vasculitis in these patients.

Rheumatology, vol. 9, pp. 174–189 (Karger, Basel 1986)

Methotrexate and Placebo Therapy in Rheumatoid Arthritis[1]

Cooperative Systematic Studies of Rheumatic Diseases Program:

H. James Williams, Department of Internal Medicine, University of Utah School of Medicine; *John R. Ward,* Department of Internal Medicine, University of Utah School of Medicine; *Cecil O. Samuelson, Jr.,* Department of Internal Medicine, University of Utah School of Medicine; *Robert F. Willkens,* Division of Arthritis, Harborview Medical Center, Seattle; *Graciela S. Alarcón,* Division of Clinical Immunology and Rheumatology, University of Alabama in Birmingham; *Maria Guttadauria,* Department of Medicine, State University of New York, Downstate Medical Center, Brooklyn; *Cheryl Yarboro,* National Institute of Arthritis, Diabetes, Digestive and Kidney Diseases, Bethesda; *Richard P. Polisson,* Division of Rheumatic and Genetic Diseases, Duke University Medical Center, Durham; *Lillian Szydlo,* Division of Rheumatology, School of Medicine, University of California, Los Angeles; *Michael E. Luggen,* Division of Immunology, University of Cincinnati Medical Center, Cincinnati; *Lynn M. Billingsley,* Rheumatology Division, Johns Hopkins University at Good Samaritan Hospital, Baltimore; *Stephen L. Dahl,* Department of Internal Medicine, University of Utah School of Medicine

Immunosuppressive therapy for rheumatoid arthritis (RA) resistant to conventional therapies is controversial and hazardous. Methotrexate (MTX) has been used to successfully treat several rheumatic disorders such as psoriatic arthritis [1, 2], dermatomyositis [3], and RA. However, MTX treatment for RA is based on uncontrolled studies [4–6]. It has been demonstrated to have anti-inflammatory action [7]. Earlier reported toxicity with the production of hepatic fibrosis has been modified by better patient selection and dosage manipulation. To evaluate the use of MTX in treating RA, a multicenter double-blind controlled study was initiated to compare MTX and placebo in patients with definite or classical RA. The study was undertaken by the Cooperative Systematic Studies of Rheumatic Diseases, consisting of

[1] Supported by NIADDK contract NO1 AM-1-2204 and Public Health Service research grant RR 00068-22 from the Division of Research Resources.

nine clinics throughout the United States[2], and funded by the National Institute of Arthritis, Diabetes, and Digestive and Kidney Diseases. Detailed results of this trial have been reported [8]. This report will summarize the initial report and analyze the effect of the weekly MTX dose on the response variables.

Methods

Patient Selection Criteria

Patients were 18 years of age or older and had definite or classical RA by the American Rheumatism Association criteria [9] of more than 6 months duration with onset after age 16. The arthritis was incompletely controlled with conventional individualized doses of salicylates or nonsteroidal anti-inflammatory drugs (NSAID) [8]. Patients had not received antimalarials, chrysotherapy, *D*-penicillamine, cytotoxic drugs, or pyrazalon derivatives for the preceding 2 months. Women patients and sexual partners of male patients had no childbearing potential or were practicing successful contraception for at least three months before entry into the study.

To be eligible, no patient could have received prior treatment with MTX. All had received previous treatment with either parenteral gold (total dose of at least 750 mg or development of toxicity) or *D*-penicillamine ($>$ 500 mg/day for 3 months or development of toxicity).

All patients had six or more swollen joints considered capable of responding to drug therapy and at least two of the following: (1) nine or more tender joints; (2) 45 min or more of morning stiffness, and (3) Westergren erythrocyte sedimentation rate (ESR) of 28 mm/h or greater.

Exclusion criteria have been published [8] and included overt obesity, insulin-dependent diabetes, and alcoholic intake of more than 14 oz of 100 proof liquor or equivalent per week.

Study Design

The study was an 18-week double-blind parallel trial comparing MTX and placebo. MTX in 2.5-mg tablets and placebo tablets were indistinguishable. Study medication was given as one table every 12 h for three consecutive doses each week or a total

[2] The participating clinics and their directors represent eight university-based centers and the National Institutes of Health as follows: Duke University, *Richard P. Polisson;* Johns Hopkins University, *Lynn Billingsley;* State University of New York, Downstate Medical Center, *David Kaplan;* University of Alabama in Birmingham, *Graciela S. Alarcón;* University of California, Los Angeles, *Harold E. Paulus;* University of Cincinnati, *Michael E. Luggen;* University of Utah, *Cecil O. Samuelson, Jr.;* University of Washington, Harborview Medical Center, *Robert F. Willkens;* National Institute of Arthritis, Diabetes, Digestive and Kidney Diseases, Arthritis and Rheumatism Branch, *John Klippel.*

of 7.5 mg MTX per week. If the investigator judged no meaningful change in disease activity after 6 weeks of therapy, the dosage could be doubled to two tablets every 12 h for three consecutive doses each week for a total of 15 mg MTX per week.

Every patient was receiving a stable dose of aspirin and/or NSAID which was kept constant for the duration of the 18-week trial. Prednisone in doses no greater than 10 mg/day (or its equivalent) which had been stable for 1 month prior to entry and maintained throughout the trial was allowed.

Methods for Evaluation

Functional class [10] was determined upon entry into the trial. Clinical assessments of disease activity were performed at study entry and at 6, 12 and 18 weeks. The same physician examined each patient throughout the trial.

Physicians' assessments included: (1) Joint count for tenderness on pressure and/or pain on motion for 60 diarthrodial joints, graded on a scale of 0 = none, 1 = positive response on questioning, 2 = spontaneous response elicited, and 3 = withdrawal by patient on examination. (2) Joint count for swelling for 58 joints graded on a scale of 0 = none, 1 = detectable synovial thickening without loss of bony contours, 2 = loss of distinctness of bony contours, and 3 = bulging synovial proliferation with cystic characteristics. (3) Joint tenderness/pain and joint swelling scores were derived by summing the severity scores for each joint. (4) Grip strength using a mercury column sphygmomanometer with a standard grip bag [11]. (5) 50 foot walk time [11]. (6) Circumference in millimeters of proximal interphalangeal (PIP) joints of the 4 fingers of each hand, as measured by an arthrocircameter [12]. The sum of circumferences of the joints measured is divided by the number of joints to obtain an average circumference of PIP joints for each patient. (7) Physician assessment of disease activity on a scale of 1 = asymptomatic, 2 = mild, 3 = moderate, 4 = severe, or 5 = very severe.

Patient assessment included: (1) Duration of morning stiffness on the day before the visit. (2) Severity of pain on a visual analogue pain rating scale [13]. (3) Overall assessment of arthritis activity graded as 1 = asymptomatic, 2 = mild, 3 = moderate, 4 = severe, or 5 = very severe.

An important aspect of any therapy is the proportion of patients started on the therapy that obtain benefit. The major hurdle to such an analysis is defining what constitutes an adequate response. We utilized definitions for what we considered important or significant improvement for several joint count variables:

(1) Therapeutic remissions [14].This definition required 5 or more of the following for at least 2 consecutive months: morning stiffness of less than 15 min, no fatigue, no joint pain by history, no joint tenderness, no joint swelling, and Westergren ESR of less than 20 mm/h in men and less than 30 mm/h in women.

(2) 'Important improvement' for joint swelling or joint pain/tenderness. Improvement of 50% or more in the number of swollen (tender) joints, i.e. the patient showed swollen (tender) joint improvement if the number of joints improved exceeded the number of joints worsened by at least 50% of the joints initially involved and capable of response. Improved is defined as change in swelling (tenderness) to none or improvement from severe to mild. Worsening is defined as involvement of a previously uninvolved joint or a change in swelling (tenderness) to severe in a previously involved joint that was moderate or less. For example, a patient with 10 moderately swollen joints with

8 joints improved and 3 joints worsened during the trial would have significant joint improvement, but if 4 joints improved and 2 joints worsened, significant joint improvement would not have occurred.

(3) Important improvement in patient assessment. Improvement in patient assessment by at least two grades or to grade 1.

(4) Important improvement in physician assessment. Improvement in physician assessment by at least two grades or to grade 1.

Laboratory Studies

Hemoglobin, white blood cell count (WBC), differential cell count, platelet count, total serum bilirubin, ALAT (SGOT), ASAT (SGPT), and alkaline phosphatase were performed every 3 weeks. Patients were withdrawn if any of the following abnormalities occurred: (1) WBC $< 3{,}000/mm^3$ or absolute PMNs $< 1{,}500/mm^3$; (2) platelet count $< 100{,}000/mm^3$; (3) twice the upper limit of normal range for total bilirubin, ALAT, ASAT, or alkaline phosphatase. Westergren ESR was done every 6 weeks and rheumatoid factor at entry and completion.

Adverse Reactions

Patients were interviewed every 3 weeks for new symptoms and the physician expressed an opinion concerning their relationship to the test medication. Only those reactions leading to withdrawal or considered as probably or definitely drug related are reported. All patients who were withdrawn for adverse reactions were followed until resolution.

Compliance

Every patient was questioned at each visit to determine adherence to their medication dose for aspirin, NSAIDs, glucocorticoids, analgesics, and test medications. In addition, pill counts of the test drugs were done.

Statistical Analysis

Statistical analyses were done to characterize the effects of MTX in all patients who were entered, and to characterize improvement in that group of patients who were able to continue treatment until completion of the study. Measures of toxicity were also described and tested. Due to the non-normal distributions of many of the response variables, statistical tests of demographics, baseline values, and improvement from first to last visit were made by Kruskal-Wallis and Mann-Whitney tests for continuous variables, and Pearson's and/or Mantel-Haenszel chi-square tests for categorical outcomes. Analysis of covariance controlling for baseline values was performed when distributional assumptions were appropriate. Two-sided tests of significance at $\alpha = 0.05$ are reported. Pairwise comparison of treatment groups controlled the probability of type I Error by using $\alpha = 0.017$.

Improvement over the duration of the study was plotted for key response variables. Statistical tests at each time point were performed at $\alpha = 0.01$ to control at less than one the expected number of spuriously significant results resulting from multiple comparisons.

Table I. Description of 189 patients entering the study by treatment group

Variable	Placebo (n = 94)	Methotrexate (n = 95)
Mean age	55.0	52.8
% female	71	71
Race (white, black, other)	60/18/16*	76/11/8
Physician assessment of disease severity		
Mild	0	1%
Moderate	43%	52%
Severe	57%	47%

* p = 0.044.

Results

Patients Studied

One hundred eighty-nine patients were admitted to the trial and randomly assigned to one of the treatment groups. Eighty patients were withdrawn from the trial including 1 patient who was withdrawn at the final visit. This resulted in two groups for analysis: (1) those who began the trial (189 patients), and (2) those who completed the 18-week trial (110 patients). The latter group included the patient withdrawn at the final visit.

Response to Treatment for All Patients Who Entered the Trial

Except for a greater number of caucasian patients in the MTX group, there were no significant differences at entry for demographic and selected RA variables between the 95 patients in the MTX group and the 94 patients on placebo (table I). The response to treatment of all patients was analyzed using the definitions previously described (see 'Methods'). No remissions were noted after 18 weeks of therapy. A significant proportion of patients on MTX achieved 'important improvement' in pain/tenderness and joint swelling scores when compared to patients receiving placebo (table II). Thirty (32%) of the patients on MTX therapy experienced this degree of improvement for joint pain/tenderness and 20 of the patients (21%) had similar improvement in joint swelling. Both of these results were statistically su-

Table II. Proportion of patients with 'important improvement' by treatment group for all patients who entered the trial

Variable	Placebo, %	MTX, %	p value
Improvement[1]			
Pain/tenderness[2]	11	32	0.001
Swelling[2]	4	21	0.002
Patient assessment[3]	7	11	0.627
Physician assessment[2]	3	10	0.163

[1] See text for definition of 'important improvement'.
[2] n = 94 MTX, 89 placebo. Others omitted because of change in observer.
[3] n = 95 MTX, 94 placebo.

perior to the placebo group where 'important improvement' was seen in 10 patients (11%) for joint pain/tenderness and 4 patients (4%) for joint swelling. However, the number of patients who achieved 'important improvement' in patient or physician assessment of disease severity was similar for MTX and placebo treatments.

Response to Treatment for Patients Who Completed 18 Weeks in the Trial

One hundred ten patients remained in the trial at completion including the patient withdrawn at the final visit. There were 57 in the MTX group and 53 in the placebo group. No differences between the groups in demographic characteristics, including race, and disease duration and severity at entry were noted.

Patients receiving MTX were significantly better than patients given placebo for all clinical variables measured (table III). The number of tender joints decreased with a median difference of 10 in MTX-treated patients and with a median difference of 2 in placebo-treated patients. Median joint pain/tenderness scores decreased by 19 in MTX patients and by 5 in placebo-treated patients. The number of swollen joints and the swollen joint score also showed statistically significant improvement in MTX patients as opposed to placebo-treated patients. Similarly, there was significant improvement in duration of morning stiffness, grip strength, average of PIP joint circumference for each hand, walk time, pain perception by the patient, and patient and physician assessment of disease activity.

Table III. Median (mean) changes[1] in clinical variables between entry and conclusion of trial in 110 patients completing 18 weeks of therapy by treatment groups

	Placebo	Methotrexate	p value
Morning stiffness, min	60 (88)	105 (117)	0.037[2]
Grip strength, mm/Hg			
Right	–1 (2)	23 (32)	<0.0001[2]
Left	0 (0)	23 (34)	<0.0001[2]
PIP circumference, mm			
Right hand	0 (0)	1 (1)	0.036[3]
Left hand	0 (–1)	1 (1)	0.009[3]
Walking time, s	1 (2)	2 (4)	0.026[2]
Pain-analogue scale, %	0 (6)	22 (23)	<0.001[3]
Patient assessment (scale 1–5)	0 (0)	1 (1)	0.004[4]
Physician assessment (scale 1–5)	0 (0)	1 (1)	0.0002[4]
Joint pain/tenderness count	2 (3)	10 (11)	<0.001[3]
Joint swelling count	–1 (–1)	6 (7)	<0.001[3]
Joint pain/tenderness score	5 (8)	19 (21)	<0.001[2]
Joint swelling score	–2 (0)	11 (13)	<0.001[2]

[1] Improvement indicated by a positive number.
[2] Mann-Whitney analysis.
[3] Analysis of co-variance.
[4] Mantel-Haenszel chi-square.

Effects on extraarticular features including subcutaneous nodules, salivary gland enlargement, vasculitis, episcleritis and scleromalacia, keratoconjunctivitis sicca syndrome, pleuritis, pulmonary lesions, and pericarditis were studied. Because the number of patients with these features was small and no effect of either treatment was apparent except for subcutaneous nodules, no formal analysis is provided except for rheumatoid nodules. Five patients on MTX developed new nodules during the period of treatment while 10 patients had a decrease in the number of nodules. This was significantly better ($p = 0.008$) than the placebo group where 12 patients developed new nodules and three patients had a reduction in the number of nodules.

Changes in hemoglobin, platelet count, and ESR were analyzed for all patients who had an unequivocal abnormality at entry (hemoglobin ‹ 12 g/dl, platelet count › 450,000/mm^3, ESR › 28/mm/h). The median increase in hemoglobin for MTX was 1.0 g/dl as opposed to 0.4 for placebo patients (table IV). The median decrease of

Table IV. Median (mean) differences[1] in selected laboratory variables between entry and conclusion of trial for patients completing 18 weeks by treatment group

Variable	Placebo	MTX	p value[2]
Hemoglobin, g/dl (< 12 at visit 1)	0.4 (0.3)	1.0 (1.1)	0.030
Platelets, X $10^3/mm^3$ (> 450 at visit 1)	60 (64)	122 (155)	0.079
ESR, mm/h (> 28 visit 1)	1 (2)	24 (26)	<0.0001
Rheumatoid factor (inverse of titer)	0 (–3,633)	480 (1,477)	<0.0001

[1] Positive values indicate improvement.
[2] Mann-Whitney analysis.

122,000/mm^3 for platelet counts in MTX-treated patients was marginally different from the fall of 60,000/mm^3 in placebo patients. More impressive was a fall in ESR of 24 mm/h in MTX patients as opposed to 1 mm/h in placebo patients.

Thirty-nine of the 44 MTX patients and 39 of the 47 placebo patients who completed the study and had serum rheumatoid factors examined had rheumatoid factor in a titer greater than 1:40 with no significant difference in the median titer at entry. The MTX group experienced a decrease in median titer and the placebo group remained the same (table IV). None of the 44 MTX patients tested had an increase in titer by two dilutions while 20 (46%) had a decrease in titer by at least two dilutions. This contrasted to the 7 of 47 (15%) patients on placebo who had an increase in titer by at least two tube dilutions while none had a decrease in titer by two or more tube dilutions ($p < 0.00001$ by Mantel-Haenszel chi-square).

Patient Withdrawals

Eighty patients were withdrawn from the study for various reasons (table V) including the patient withdrawn at the concluding visit. Thirty of the 40 patients withdrawn for adverse drug reaction were taking MTX and accounted for 32% of all patients started on the drug. Only 3 patients on MTX were withdrawn for lack of response while 20 patients on placebo were withdrawn for this reason. Seventeen patients were withdrawn for reasons unrelated to therapy, including protocol violations, intercurrent illness, lack of cooperation, and moving from the clinic area.

Table V. Patient withdrawal by treatment group

Reason	Placebo (n = 94)	Methotrexate (n = 95)
Adverse drug reaction	10 (11%)[1]	30 (32%)
Elevated liver enzymes	4	18
Mucosal ulcers	3	5
Gastrointestinal upset	0	3
Pancytopenia	0	2
Leukopenia	0	2
Thrombocytopenia	1	0
Rash	1	0
Pulmonary infiltrate	1	0
Lack of response	20 (21%)	3 (3%)
Other	11 (12%)	6 (6%)
Total	41 (44%)	39 (41%)

[1] Percent of total number entered.

Elevated hepatic enzymes was the most common untoward drug effect in both groups with 19% of all MTX patients being withdrawn for liver enzyme levels exceeding twice the upper limit of normal. This compared to 4% of the placebo group. Oral and vaginal ulcers occurred in both treatment groups, but withdrawal for gastrointestinal upset was only noted in patients on MTX. Two patients taking MTX developed pancytopenia and two others developed leukopenia. One patient on placebo developed thrombocytopenia with epistaxis. Rash and a pulmonary infiltrate each resulted in withdrawal of one patient receiving placebo. All adverse reactions resolved following cessation of therapy and there were no deaths in the trial.

7.5 mg/Week vs. 15 mg/Week in Patients Completing the Trial

Nineteen patients (33% of patients completing 18 weeks) of the MTX group remained on 7.5 mg MTX per week while the remainder increased the dose to 15 mg per week after 6 weeks of therapy. Seven patients (13%) in the placebo group remained on three tablets a week while 46 (87%) doubled the number after 6 weeks. Of the 38 MTX patients who increased the weekly dose, 32 increased the dose after 6 weeks, 2 increased at 9 weeks and 4 increased at 12 weeks. At entry to

Table VI. Median (mean) differences[1] in selected clinical variables between entry and conclusion of the trial in patients on MTX who completed trial by maximum weekly dose

Variable	7.5 mg/week (n = 19)	15 mg/week (n = 38)	p value
Morning stiffness, min	150 (138)	83 (106)	0.167[2]
Grip strength, mm/Hg			
Right	28 (37)	19 (29)	0.232[2]
Left	22 (31)	29 (35)	0.946[2]
PIP circumference, mm			
Right	2 (3)	0 (0)	0.002[3]
Left	2 (2)	0 (0)	0.084[2]
Walking time, s	3 (3)	2 (4)	0.505[2]
Pain-analogue scale, %	30 (29)	20 (20)	0.140[3]
Patient assessment (scale 1–5)	1 (1)	1 (1)	0.493[4]
Physician assessment (scale 1–5)	1 (1)	1 (1)	0.018[4]
Joint pain/tenderness count	19 (17)	8 (8)	0.005[3]
Joint swelling count	10 (10)	6 (5)	0.046[3]
Joint pain/tenderness score	27 (34)	14 (14)	0.004[3]
Joint swelling score	17 (20)	10 (9)	0.042[2]

[1] Improvement indicated by a positive number.
[2] Mann-Whitney analysis.
[3] Analysis of co-variance.
[4] Mantel-Haenszel chi-square.

the study, there was no difference in age, sex, race, or severity or duration of disease between the group that remained on 7.5 mg MTX per week and the group that took 15 mg per week. However, 24 of the 38 patients on 15 mg MTX were taking corticosteroids as opposed to three of the 16 patients on 7.5 mg MTX. Rheumatoid factor was similar in the two groups.

Comparison of the 18-week responses for clinical variables for the two doses of MTX reveal that patients on the lower dose demonstrated greater improvement in joint pain/tenderness count and score, joint swelling count and score, physician assessment, and average PIP joint circumference of the right hand (table VI). These changes were most marked for joint pain/tenderness count and score and for the average PIP joint circumference of the right hand. The median joint count for

Table VII. Median (mean) differences[1] in selected laboratory variables between entry and conclusion of the trial in patients on MTX who completed trial by maximum weekly dose

Variable	7.5 mg/week (n = 19)	15 mg/week (n = 38)	p value[2]
Hemoglobin, g/dl (< 12 at visit 1)	1 (1)	1 (1)	0.786
Platelets, X $10^3/mm^3$ (> 450 at visit 1)	128 (118)	122 (174)	0.734
ESR, mm/h (>28 at visit 1)	37 (35)	22 (20)	0.034
Rheumatoid factor (inverse of titer)	640 (2,484)	300 (956)	0.377

[1] Improvement indicated by a positive number.
[2] Mann-Whitney analysis.

Table VIII. MTX patient withdrawals by maximum weekly dose

Reason	7.5 mg/week (n = 35)	15 mg/week (n = 60)
Adverse drug reaction	14	16
Elevated hepatic enzymes	8	10
Mucosal ulcers	4	1
Pancytopenia	0	2
Leukopenia	1	1
Gastrointestinal distress	1	2
Lack of response	0	3
Other	3	3
Total withdrawals	17	22

pain/tenderness fell by 19 in patients on 7.5 mg/week and the joint pain/tenderness score fell by 27. Counts and scores for pain/tenderness fell by 8 and 14, respectively, in the 15-mg/week group.

The only significant difference in laboratory variables was seen in ESR (table VII). The ESR fell by a median value of 37 mm/h in the lower dose group compared to 22 mm/h in higher dose group.

Evaluating the two doses at 6 weeks for 'important improvement' in joint counts indicates that 42% of the patients that remained on 7.5

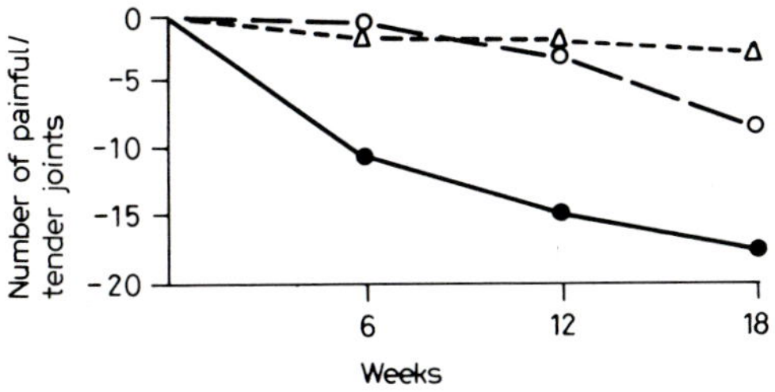

Fig. 1. Response by pain/tenderness joint count over time for placebo (△), 7.5 mg/week MTX, (●), and 15 mg/week MTX (○).

mg/week throughout the 18-week study experienced this degree of improvement for pain/tenderness and for swelling compared to 8% for pain/tenderness and 6% for swelling in patients whose dosage was increased. By the 18th week, 'important improvement' as assessed by joint pain/tenderness and swelling increased to 44% and 28%, respectively, in the 15-mg/week group. However, at the end of the study, 74% and 53% of the lower dose group met the criteria for 'important improvement' in joint pain/tenderness and swelling, respectively.

An evaluation of the time to response in patients completing 18 weeks of therapy indicates that improvement begins at about 6 weeks in the 7.5-mg/week group and is more striking than in the 15-mg/week group. An example using joint pain/tenderness counts is shown in figure 1.

Withdrawals for adverse drug effect were similar in the two groups (table VIII). Mucosal ulcerations were more common on the lower dose while pancytopenia occurred only on the higher dose. Leukopenia occurred in 1 patient on each dose.

Discussion

A detailed analysis of this study has previously been reported [8]. There are seven reports of other trials of MTX in RA patients [4–6, 15–18]. These studies are generally small and uncontrolled. All studies defined efficacy and drug response differently but all concluded that MTX was effective in their patients with previously unresponsive arthritis.

The dosage schedule selected for our study was based on the experience in treating psoriasis with MTX. Early treatment schedules had included small daily doses of MTX with or without short rest periods, large parenteral doses, or large weekly oral doses [19]. *Weinstein and Frost* [20] suggested another regimen of three doses a week at 12-hour intervals to provide a pharmacologic effect for the 37-hour period of rapid psoriatic cell turnover. This method also yielded lower peak serum levels and was therefore potentially less toxic. This manner of administration became the preferred method of most dermatologists [21] and continues to be one of the two recommended dosage schedules for treatment of psoriasis [22]. This dosage has been adopted by most of the investigators studying MTX in RA. However, for the treatment of psoriasis there appears to be little difference between this regimen and giving the dose of MTX as a single weekly dose [22].

The results of this double-blind trial are in agreement with the unblinded studies previously published. However, marked improvement as defined by our definitions for 'important improvement' was seen in 32% of all patients entered into the trial for joint pain/tenderness and 21% for joint swelling. The withdrawal rate of 32% for adverse drug reactions in this trial is higher than the experience derived from previously published unblinded studies of 22%.

The effect of MTX dosage on treatment outcomes has not previously been analyzed. In this study, all patients were started on 7.5 mg MTX per week and physicians had the option of doubling the dose after the sixth week. Increasing the dose was to be done for any patients who were not responding. Three clinics increased this dosage in all patients (both placebo and MTX groups) to 6 tablets a week. Whether all patients were unresponsive or this is an indication of some unconscious bias is unknown. Only seven placebo-treated patients were continued on 3 tablets a week suggesting that most of the placebo patients were unresponsive. One-third of the MTX-treated patients who completed the study remained on the lower dose. The number of patients on prednisone in the higher dose group suggested that these patients had more severe disease. However, analysis of duration and initial severity of disease failed to demonstrate any significant differences between the two dosage groups.

Patients who completed the study on 7.5 mg/week generally responded sooner and to a greater degree than patients whose doses were doubled. Increasing the dose in patients who were not responsive led to improvement in some patients but improvement in this group was not as marked as that seen in the group responding to 7.5 mg/week at 6 weeks (fig. 1).

Withdrawals did not seem to be any greater with the higher dose with the possible exception of hematological suppression. Elevated liver enzymes occurred equally on both doses. Mucosal ulcerations occurred more frequently on the lower dose but seem to be a function of time rather than dose. Four patients had evidence of bone marrow suppression, 2 with leukopenia and 2 with pancytopenia. One of these 4 patients developed leukopenia on the lower dose but the other 3 were taking the higher dose. The numbers are small but may indicate a potential for greater marrow toxicity with higher doses of MTX.

Several questions remain unanswered. Would the response seen in the 15-mg/week patients be the same if the MTX dosage had remained at 7.5-mg/week? Was the later response a function of time as opposed to dose? Would the response of the 7.5-mg/week group be even greater if the dosage had been doubled? The optimal dose of MTX is still undetermined as is the optimal dosage schedule.

Conclusions

The results of this 18-week trial reveal that MTX is superior to placebo in all variables measured. However, toxicity to MTX was great and one-third of the patients on MTX were withdrawn for adverse drug effects. Additional analysis indicates that patients begin responding by 6 weeks and respond to 7.5 mg/week. If this dose does not cause significant improvement, doubling the dose will help some but not many. When response in the lower doses is unsatisfactory, improvement on 15 mg/week is less impressive than that observed in the patients who respond early to 7.5 mg/week.

Acknowledgements

The assistance of the staff at the Coordinating Center was most valuable. Specific important contributions were made by *Miki L. Coleman*, MS, applied biostatistician, and *Wanda L. Moore*, administrative assistant.

References

1 Black, R.L.; O'Brien W.M.; Van Scott, E.J.; Auerbach, R.; Eisen, A.Z.; Bunim, J.J.: Methotrexate therapy in psoriatic arthritis. J. Am. med. Ass. *189:* 743–747 (1964).

2 Willkens, R.F.; Williams, H.J.; Ward, J.R.; Egger, M.J.; Reading, J.C.; Clements, P.J.; Cathcart, E.S.; Samuelson, C.O.; Jr.; Solsky, M.A.; Kaplan, S.B.; Guttadauria, M.; Halla, J.T.; Weinstein, A.: Randomized, double-blind, placebo controlled trial of low dose pulse methotrexate in psoriatic arthritis. Arthritis Rheum. *27:* 376–381 (1984).

3 Arnett, F.C.; Whelton, J.C.; Zizic, T.M.; Stevens, M.B.: Methotrexate therapy in polymyositis. Ann. rheum. Dis. *32:* 536–546 (1973).

4 Gubner, R.; August, S.; Ginsburg, V.: Therapeutic suppression of tissue reactivity. II. Effects of aminopterin in rheumatoid arthritis and psoriasis. Am. J. med. Sci. *221:* 176–182 (1951).

5 Willkens, R.F.; Watson, M.A..; Paxson, C.S.: Low dose pulse methotrexate therapy of rheumatoid arthritis. J. Rheumatol. *7:* 501–505 (1980).

6 Wilke, W.S.; Calabrese, L.H.; Scherbel, A.L.: Methotrexate in the treatment of rheumatoid arthritis. Cleve. Clin. Q. *47:* 305–309 (1980).

7 Hurd, E.R.: Immunosuppressive and antiinflammatory properties of cyclophosphamide, azathioprine, and methotrexate. Arthritis Rheum. *16:* 84–88 (1973).

8 Williams, H.J.; Willkens, R.F.; Samuelson, C.O..; Jr.; Alarcón, G.S.; Guttadauria, M.; Yarboro, C.; Polisson, R.P.; Weiner, S.R.; Luggen, M.E.; Billingsley, L.M.; Dahl, S.L..; Egger, M.J.; Reading, J.C.; Ward, JR.: Comparison of low dose oral pulse methotrexate and placebo in the treatment of rheumatoid arthritis. A controlled clinical trial. Arthritis Rheum. *28:* 721–730 (1985).

9 Ropes, M.W.; Bennett, G.A.; Cobb, S.; Jacox, R.; Jessar, R.A.: 1958 revision of diagnostic criteria for rheumatoid arthritis. Bull, rheum. Dis. *9:* 175–176 (1958).

10 Steinbrocker, O.; Traeger, C.H.; Batterman, R.C.; Therapeutic criteria in rheumatoid arthritis. J. Am. med. Ass. *140:* 659–662 (1949).

11 The Cooperating Clinics Committee of the American Rheumatism Association: A seven-day variability study of 499 patients with peripheral rheumatoid arthritis. Arthritis Rheum. *8:* 302–334 (1965).

12 Willkens, R.F.; Gleichert, J.E.; Gade, E.T.: Proximal interphalangeal joint measurement by arthrocircameter. Ann. rheum. Dis. *32:* 585–586 (1973).

13 Huskisson, E.C.: Assessment for clinical trials. Clin. rheum. Dis. *2:* 37–49 (1976).

14 Pinals, R.S.; Masi, A.T.; Larsen, R.A.: Subcommittee for Criteria of Remission in Rheumatoid Arthritis of the American Rheumatism Association Diagnostic and Therapeutic Criteria Committee: Preliminary criteria for clinical remission in rheumatoid arthritis. Arthritis Rheum. *24:* 1308–1315 (1981).

15 Michaels, R.M.; Nashel, D.J.; Leonard, A.; Sliwinski, A.J.; Derbes, S.J.: Weekly intravenous methotrexate in the treatment of rheumatoid arthritis. Arthritis Rheum. *25:* 339–341 (1982).

16 Steinsson, K.; Weinstein, A.; Korn, J.; Abeles, M.: Low dose methotrexate in rheumatoid arthritis. J. Rheumatol. *9:* 860–866 (1982).

17 Groff, G.D.; Shenberger, K.N.; Wilke, W.S.; Taylor, T.H.: Low dose oral methotrexate in rheumatoid arthritis. An uncontrolled trial and review of the literature. Semin. Arthritis. Rheumatism *12:* 333–347 (1983).
18 Hoffmeister, R.T.: Methotrexate therapy in rheumatoid arthritis. 15 years experience. Am. J. med. Proc. Symp. Oral Gold Therapy in Rheumatoid Arthritis, 1983, pp. 69–73.
19 Weinstein, G.D.: Methotrexate. Ann. intern. Med. *86:* 199–204 (1977).
20 Weinstein, G.D.; Frost, P.: Methotrexate for psoriasis.A new therapeutic schedule. Archs Derm. *103:* 33–38 (1971).
21 Bergstresser, P.R.; Schreiber, S.H.; Weinstein, G.D.: Systemic chemotherapy for psoriasis. A national survey. Archs Derm. *112:* 977–981 (1976).
22 Roenigk, H.H., Jr.; Auerbach, R.; Maibach, H.I.; Weinstein, G.D.: Methotrexate guidelines – revised. J. Am. Acad.Derm. *6:* 145–155 (1982).

H. James Williams, MD, University of Utah Medical Center,
50 North Medical Drive, Salt Lake City, UT 84132 (USA)

Rheumatology, vol. 9, pp. 190–191 (Karger, Basel 1986)

Discussion 8

Wilke: We are open for discussion now. Let me start the ball rolling. Dr. *Williams*, in light of the fact that in your study, the patient evaluation, patients' own impression whether or not they were getting better, was not 'statistically significant'. However, the analogue scale did show significance. Try to explain that to me, and then tell me whether the study was clinically significant.

Williams: The answer to the second one is easier than the answer of the first one. Was the study clinically significant? We've done several studies now, comparing different second- and third-line agents. And this has probably shown the greatest difference between the placebo group and the active medication group. I think we are talking in numbers of joints different. That is significant. A median change of 10 joints in the treated to 2 joints in the placebo patients is, I think, significant clinical difference, not just a statistical difference. How do you explain the fact that the patients' assessment was not significant or was not as significant as their assessement of pain. The analogue scale is a much more sensitive scale than a 4-point scale of none, mild, moderate, severe. And the changes were small enough there that there was no clinical significance. However, on the percentage scale it did show that there was a difference. It is interesting on another pain scale, which was mild, moderate, severe, that the differences were not as clear as the differences on the analogue scale, and I think it's the instrument you're using.

Wilke: I'll ask you one more question and in the same line. Were you one of the major workers in this study?

Williams: Correct.

Wilke: You got a chance to see a fair number of patients at your institution, is that correct?

Williams: Yes, the University of Utah is one of the biggest clinics, and I saw a fair number of patients.

Wilke: Do you think you could predict, when you saw the patients, whether they were taking methotrexate or not?

Williams: We did not do that in this study. In other studies we have asked the people to predict...

Wilke: No, I didn't ask you that... I just said, do you think, you could tell, and were you right when you thought you could? I could tell in my study. And I was right most of the time.

Williams: When we broke the code, it turned out that every one of my patients was on methotrexate. And every one of Dr. *Ward's* patients was on placebo. I thought it was a great trial. So that I could not say whether or not I could predict, because all of my patients did well.

Wilke: Fair enough.

Rau: I have a question for Dr. *Williams*. You had 2 groups of patients. One group responded to 7.5 mg very well, and the second group didn't respond or responded only to 15 mg, but not as well as the first group. Do you find any difference between these two groups in respect of the stage of the disease, the duration of the disease, the joint destruction. That means, are there more severe cases on the 15 mg dose?

Williams: Let's make sure we understand that both groups started out on 7.5 mg, and they didn't divide until 6 weeks. So, those who are unresponsive to methotrexate ended in the 15-mg group. But in comparing the two groups at 6 weeks, there were no differences between the groups, except that the group on 15 mg a day had more patients who were taking prednisone in a significant difference. I can look up the numbers, I don't remember what they are, suggesting that perhaps they were a more severe group. However, if you look at severity of disease as assessed by the physician, there were no differences between the groups. But there were more patients on 15 mg taking prednisone than those who remained on 7.5. That was the only suggestion of any difference between the 2 groups.

Rau: You know I don't mean the activity of the joints or of the disease, but the severity of the joint destruction. For if you have an advanced joint destruction, then you cannot expect to get a good therapeutic effect on these joints, for every movement is painful and causes an inflammation of the synovial membrane in this case of a so-called secondary osteoarthritis. So you will have pain and you will have inflammation after movement, and you can't influence this by drugs.

Williams: We measured this by (1) the physician's assessement of the severity of the disease; (2) functional class, and (3) anatomic class, and there were no differences between the two groups.

Furst: Dr. *Williams,* did you do baseline X-rays?

Williams: It's only an 18-week trial, we didn't do X-rays.

Weinstein: I have a question for Dr. *Williams* relating to the same point that Dr. *Rau* asked. Do you think that if you actually had controlled that part of the study, that 6-week point, and had put those patients, taken half those patients who were resistant and put them on 15 mg, and the other half left on 7.5 mg, do you think you would have seen the same degree of responsiveness in those patients who have been left on the 7.5 mg, and it was just a matter of time that some patients don't respond in 6 weeks?

Williams: In the manuscript we discussed that very thing, pointing out that those patients who showed the most improvement on 15 mg may well have responded if we had left them on 7.5 mg. We don't know that. We don't expect that they would have been any better if we had left them on 7.5 mg. But there is the possibility that they would have been the same. We also can't say that the group who responded to 7.5 mg wouldn't have had a greater response if we had increased it to 15 mg. We wouldn't have expected them to have a lesser response. So, we can't say what the 2 groups would have done if we had reversed them.

Wilke: Was there any, well, one of the problems that I think everyone is alluding to, is the possibility that some of the patients had fixed deformities that may have been a little difficult to show improvement in certain things?

Williams: We had a definition of what consisted of an active joint. And I concluded up there that they had to be swollen and so many tender and so forth. But the joint had to be responsive in terms of the fact that if there were fixed deformities it could not be included in the trial and could not be evaluated. If there had been joint replacements, we could not include that joint. They were allowed one intra-articular injection and that joint was excluded from analysis, so that joints that were included for analysis were responsive, active joints.

BE

Rheumatology, vol. 9, pp. 192–212 (Karger, Basel 1986)

Methotrexate Hepatotoxicity in Psoriasis and Psoriatic Arthritis

A Review

A. Nyfors

Department of Dermatology, Haukeland University Hospital, Bergen, Norway

Introduction

It would be useful to have read the previous review on methotrexate (MTX) therapy in psoriasis and psoriatic arthritis [37] before reading this review. In this paper I shall ask 11 questions concerning the hepatotoxicity of MTX in psoriasis, go through the available documentation (where psoriasis and psoriatic arthritis have not been separated as patients with arthritis have had psoriasis of the skin) of this toxicity, and try to give answers to the questions. The readers may wish that other questions had been asked and answered, too. The chosen questions below are natural questions on behalf of the patients, whom we are responsible for the care of. For the reader with less interest in the details, it may be enough to read the questions and the answers.

The History of MTX Hepatotoxicity

Pathological findings in liver biopsies after MTX therapy for leukemia have been reported by several authors. In 1955, *Colsky* et al. [7] described fibrosis in postmortem liver specimens of 3 MTX-treated leukemic children. The cumulative dose of MTX varied from 237 to 1,057 mg over a 6–9-month' treatment period. MTX was given periodically in daily doses of 2,5–10 mg. *Hutter* et al. [19] in 1960 found no fibrosis in 18 'normal' children who died suddenly, but fibrosis in 31% of 26 untreated leukemic children and fibrosis in 80% of 247 leukemic children treated with cytostatic drugs such as MTX or prednisone. The hepatotoxicity of MTX in the treatment of leukemics was obvious already in 1960. It lasted until 1969 before biopsy-proven liver damage of MTX in psoriatics was systematically investigated. In 1964, *O'Rourke and Eckert* [38] published a case report of MTX-induced hepatic injury in an adult with psoriasis. In 1968, *Coe and Bull* [6] reported 3 cases of cirrhosis associated with MTX treatment of psoriasis. Further similar reports of liver biopsy verified hepatotoxicity from MTX therapy in psoriatics were published during 1969 [11, 29]. The findings inspired dermatologists to systematic studies of MTX-treated psoriatics by liver biopsies and other means.

Questions

At the beginning of these prospective studies [36], a series of questions were asked.

(1) How and how often is the liver histology pathological in the so-called normal population over the age of 40 years (psoriatics over this age is the main indication for MTX therapy of psoriasis)?

(2) Is liver histology normal in psoriatics not treated with cytostatics?

(3) What are the quantitative and qualitative findings of liver pathology in psoriatics after long-term MTX therapy?

(4) What is the morphogenic development of cirrhosis during MTX therapy?

(5) What is the course and prognosis of the developed cirrhosis in MTX-treated psoriatics?

(6) Which main factors influence liver histology in psoriatics?

(7) What is the best control investigation to detect possible liver damage in long-term MTX-treated psoriatics?

(8) Do MTX dosage schedules influence the hepatotoxicity of MTX therapy?

(9) What can be done to prevent or minimize liver damage during MTX therapy for psoriasis?

(10) Interpretations of liver biopsy findings?

(11) How should the patients be cared for in the future?

To answer these questions I shall go through the different studies in the field as well as comment on MTX drug interactions and the influence of other drugs upon the liver.

Studies of MTX Hepatotoxicity in Psoriatics

Findings in Pre-MTX Liver Biopsies from Psoriatics

In 1968 and 1969 – as mentioned above – a few cases of cirrhosis in patients with psoriasis on MTX therapy were reported. Among these psoriatics 5 had been subjected to a post-MTX liver biopsy, while there was no histological evaluation at all of the liver tissue prior to MTX therapy [6, 27]. A normal pre-MTX liver biopsy would have indicated, though not proved, that the cirrhosis was due to the MTX treatment. In 1969, therefore, the author started taking pre-MTX liver biopsies in psoriatics for whom this treatment was indicated [36]. This posed a new problem: Are liver biopsies from psoriatics normal or are there special pathological changes in this group of patients? The next question was naturally: What is the state of the liver tissue in an age-matched group of the normal population? Neither question could be answered in 1969. Since that time the former question has been partly answered by the numerous liver biopsies from psoriatics prior to the institution of MTX therapy [see table V in ref. 29], while – for ethical reasons – there have not been major studies on liver biopsies in healthy controls. But there have been autopsy studies, two of which will be mentioned below. Among 60 patients who had died from Parkinson's disease, heart disease, or in traffic accidents *Zachariae and Søgaard* [59] in 1973 found a pathological liver histology in 40 (67%): 5 controls had liver cirrhosis (3%) and 12 controls had fibrosis (7%). The mean age was 61 years. *Hilden* et al. [15] in 1977 found in liver specimens from 503 persons killed in traffic accidents, mean age 45 years: fatty change in 24%, nonspecific portal inflammation in 1.4%, alcoholic hepatitis in 1.2%, portal fibrosis in 1%, and cirrhosis in 0%.

Nyfors and Poulsen [29] in 1976 found pathological liver histology in 51% of 123 consecutive liver biopsies from 123 psoriatics prior to institution of MTX therapy. The mean age was 52 years. The pathological changes were mainly fatty change and/or nonspecific reactive hepatitis. Portal fibrosis was present in 1 patient and cirrhosis in 1, which makes 1% (1 out of 116 successful liver biopsies) [see table III in ref. 29]. Among 200 pre-MTX liver biopsies from psoriatics, mean age 47 years, *Weinstein* et al. [58] in 1973 found 66% pathological liver biopsies, including 23% with fibrosis and 1.5% with cirrhosis. On the basis of the literature, these authors reported a 3% prevalence of cirrhosis in the corresponding age group of the normal population. *Zachariae and Søgaard* [59] in 1973 found pathological changes in two-thirds: cirrhosis in 0%, and fibrosis in 6% of pre-MTX liver biopsies from 47 psoriatics, mean age 54 years. Perusal of the literature in 1975 [table V in ref. 29] showed that pathological changes had been found in an average of 68% of a total of 399 pre-MTX liver biopsies from psoriatics, whereas the mean prevalence of fibrosis and cirrhosis was 11 and 1%, respectively. The mean age of these 477 patients was 49 years.

Pre- and Post-MTX Liver Biopsies in Psoriatics

As for materials of pre- and post-MTX liver biopsies, from the same psoriatic patient, only a few studies with a major number of liver biopsies are on record. This is all the more regrettable as series of liver biopsies from the same patient are best suited for elucidating hepatic changes arising during the MTX treatment of psoriasis. Among 88 psoriatics having 1 pre-MTX and at least 1 post-MTX biopsy *Nyfors and Poulsen* [30] in 1976 found that cirrhosis developed in 5 and fibrosis in 6 after a mean cumulative dose of 1,733 mg: Development of cirrhosis and fibrosis was not significantly more common ($p = 0.062$) in psoriatics having abnormal pre-MTX biopsy than in patients having a normal pre-MTX biopsy. By a discriminant analysis of the clinical data it was not possible to demonstrate a single factor as being responsible for the development of cirrhosis/fibrosis. It seemed to be multifactorial. In the named study a risk index was set up, comprising the cumulative dose of MTX, age, obesity, admitted intake of alcohol during MTX therapy, and chief histological diagnosis on the pre-MTX liver biopsy. The risk index proved applicable for predicting which patients would be most likely to develop cirrhosis or fibrosis during treatment of psoriasis with weekly oral doses.

Weinstein et al. [58] in 1973, among 38 MTX treated psoriatics, each having 1 pre- and 1 post-MTX liver biopsy, found a significantly increased degree of fatty change and fibrosis ($p < 0.05$). The total cumulative dose of MTX in their study is not stated. Four cases of cirrhosis were diagnosed by the second liver biopsy. *Zachariae* et al. [61] in 1975, in a material of 56 psoriatics each having 1 pre- and 1 post-

MTX liver biopsy, found no statistically significant change in liver histology from the pre- to the post-MTX biopsy after 3–4 years of MTX treatment. In 3 less-comprehensive studies [1, 43, 57], comprising from 25 to 9 liver biopsies from psoriatics, there has been no statistically significant difference between the histology in the biopsies taken before and after MTX therapy. The explanation may be that these series are too small for statistical analysis; another possible, and more important, explanation may be that the total MTX dose has been too low for a possible hepatotoxic effect to manifest itself clearly as a single factor.

Serial Post-MTX Liver Biopsies. The practice of taking liver biopsies prior to the institution of MTX treatment of psoriatics was not introduced until the 1960s. Therefore, the literature offers a much larger number of reports on hepatic changes in post-MTX biopsies than on the development of changes from pre- to post-MTX therapy. A single post-MTX liver biopsy is of value per se in assessing the current liver status of the patient concerned. On the other hand, it does not tell us whether the changes, if any, are due to MTX as long as a morphological developmental pattern in series of liver biopsies has not been elucidated or described; nor has a single change been found to be pathognomonic of MTX.

Series of post-MTX liver biopsies from the same psoriatic have been reported, but not many. *Weinstein* et al. [58] in 1973 found no statistically significant increase in the degree or incidence of fibrosis and cirrhosis from the first to the latest liver biopsy in post-MTX liver biopsies from 81 psoriatics. The mean time interval between the first and latest post-MTX liver biopsy was 13 months. The mean MTX dose at the latest liver biopsy is not stated. There was a significant increase ($p = 0.038$) in portal inflammation in the latest biopsies. *Almeyda* et al. [1] in 1972, studying series of post-MTX liver biopsies from 30 psoriatics, observed development of fibrosis in 3 (10%) after a mean treatment period of 13 months. After MTX medication for 3–4 years, *Zachariae* et al. [61] in 1975 found development of cirrhosis in 3 cases (11%) and of fibrosis in 6 patients (21%) in series of post-MTX liver biopsies from 28 psoriatics. In contrast *Reese* et al. [43] in 1974 found no increase in the degree of fibrosis in a series of post-MTX liver biopsies from 21 psoriatics after a mean treatment period of 27 months. The conclusion reached by *Reese* and his colleagues, that liver biopsies are not needed in the control MTX treatment of psoriasis, rests on too slender a basis, both regarding the size of their material as well as the duration of the MTX medication. The authors could have suggested larger time intevals between liver biopsies as a more suitable conclusion. *Zachariae* et al. [61] in 1975 had the longest treatment period – and thus presumably also the highest mean MTX dose. And indeed, their finding of 21% of fibrosis and 11% of cirrhosis is the highest among the 4 studies mentioned above. It is but natural to expect that a hepatotoxic effect of MTX, if any, will be more strikingly manifest the higher the total dose of MTX or (presupposing no major breaks in the treatment) the longer the medication has been given.

In a series of post-MTX liver biopsies from 68 psoriatics whose mean total dose of MTX has increased from 2,918 to 3,940 mg between the first and latest post-MTX liver biopsy, *Nyfors* [32] in 1977 found 21% cirrhosis and 24% fibrosis in the latest liver biopsies compared with 13% cirrhosis and 15% fibrosis in the earliest post-MTX liver biopsies. The mean treatment period at the latest liver biopsy was 5 years and 7 months. The higher percentage of fibrosis as well as of cirrhosis in this latter material [32], treated for 5–6 years, is in agreement with the findings of *Zachariae* et al. [61] after treatment for 3–4 years.

A single post-MTX liver biopsy from psoriatics is of value primarily in assessing the liver status of a patient and thereby in deciding whether or not he/she is to continue on MTX therapy. On the other hand, a diagnosis based on a single post-MTX liver biopsy does not afford much information about the pathogenesis underlying the liver pathology, unless it is alcoholic hepatitis [5], viral hepatitis [41], or other better-elucidated histological patterns of liver disease.

Morphogenesis of Fibrosis and Cirrhosis following MTX Therapy

The author is not aware of publications which have described the pathogenesis of fibrosis and cirrhosis following MTX therapy in psoriatics. This requires histological analysis of pre-MTX as well as of several post-MTX liver biopsies from psoriatics treated with MTX for several years. However, the localization of the fibrosis and the type of cirrhosis have been reported by a number of authors. *Dahl* et al. [8] in 1971, *Roenigk* et al. [46] in 1971, *Almeyda* et al. [1] in 1972, *Palmer* [39] in 1973, *Millward-Sadler and Ryan* [26] in 1974 and *Warin* et al. [57] in 1975 have described 22 cases of cirrhosis in all, 17 of which were micronodular, and 38 cases of fibrosis, mainly portal and periportal.

Nyfors and Poulsen [34] in 1977 reported 8 psoriatics having 1 pre-MTX and at least 2 post-MTX liver biopsies each. Four developed cirrhosis and 4 fibrosis. In the cirrhotic patients the morphological findings indicated the following sequence of the changes: first multiple, small piecemeal necroses with destruction of the limiting plate, followed by stellate periportal fibrosis. Later there developed portal-portal and portal-central vein bridging, fibrosis with gradual alteration of the lobular architecture, leading ultimately to cirrhosis. The type of cirrhosis was micronodular in 3. The 4 patients with fibrosis had similar changes and a similar pattern of morphological development, but without regeneration nodules. That the above histological pattern of development in the liver biopsies in the 8 patients was due to the action of MTX is supported by the finding that the sequence was the same in all 8 patients and that a similar developmental pattern was found in another 3 studied by serial post-MTX liver biopsies. That additional pathogenic factors may play a role in the development of cirrhosis is indicated by the presence of alcoholism and central sclerosis in 2 of the above-mentioned cirrhotics. The mean period from the first liver biopsy until cirrhosis was diagnosed in the 4 patients was 36 months. While, thus, it is likely that piecemeal necrosis plays a role in the development of cirrhosis in MTX-treated psoriatics, the importance of finding of piecemeal necrosis in a single liver biopsy is uncertain. A follow-up may show none. This may be due, among other things, to sampling errors. Similar changes may occur in fibrosis.

The question whether fibrosis following MTX therapy in psoriatics is reversible cannot be answered with certainty. However, it is worth mentioning that in liver biopsies from a few (< 10) patients in whom MTX therapy was discontinued, the present author has seen a decrease of the fibrosis. The same has been observed in 3 patients despite continued MTX medication [32]. *Zachariae* et al. [62] in 1980 found in 13 patients with cirrhosis after MTX therapy for psoriasis a trend for later serial liver biopsies to be of the same rather low grading as found in the original liver biopsy, or even lower, irrespective of continued MTX therapy.

So far, we have no ideal classification of cirrhosis, and the morphogenesis of its development has not been finally elucidated [2]. The pattern of development described above indicates that cirrhosis following MTX therapy starts as many small piecemeal necroses, gradually converts into fibrous septa which ultimately may split up the lobular architecture. A partially similar morphogenesis may be observed in the progression of chronic aggressive hepatitis into cirrhosis [51]. The mild course of the cirrhosis seen in nonalcoholic and not arsenite-treated MTX-treated psoriatics [14, 62] points into the direction of a chronic persistent hepatitis more than at a chronic aggressive hepatitis.

Electron-Microscopic Findings in Liver Biopsies from Psoriatics before and after MTX Therapy

Ito Cells. In 1973, *Horvath* et al. [18] published the results of their study of Ito cells in 9 post-MTX liver biopsies from psoriatics. They found an Ito cell index (number of Ito cells per 1,000 hepatocytes) higher than normal. Their counting method is not described. According to *Horvath* et al. [18], the increase in the Ito cell index in the post-MTX liver biopsies from the 9 psoriatics was not correlated to the total dose of MTX. *Hopwood and Nyfors* [16] in 1976, point counting Ito cells in 1 pre-MTX and 1 post-MTX liver biopsy from each of 24 psoriatics, found a statistically significant increase in the number and size of the Ito cells in the post-MTX specimens ($p<0.01$). The increase in the number of the Ito cells in the post-MTX liver biopsies did not correlate significantly with the total doses of MTX. By electron microscopy it was not possible to demonstrate an essential difference in the structure of Ito cells between the pre-MTX and post-MTX liver biopsy from the same psoriatic. Whether the *increase in the number of Ito cells* during the MTX therapy is a factor in causing the fibrosis in the MTX-treated psoriatics cannot be said with certainty.

Ultrastructural Findings in Hepatocytes. Only a few ultrastructural studies of liver biopsies from MTX-treated psoriatics are on record. *Nyfors and Hopwood* [17] *in 1977 described proliferation of smooth endoplasmic reticulum* and elongated mitochondria with paracrystalline bodies and accumulation of lipid droplets. *Horvath* et al. [18] in 1973 found similar changes in the hepatocyte organelles and an increased Ito cell index. *Nyfors and Hopwood* [33] in 1977, comparing hepatocytes in pre- and post-MTX liver biopsies, found signs of cellular damage in the form of membrane whorls in the cytoplasm, an increased fat content of the cells, megamitochondria with budding, an increased number of residual bodies in the cytoplasm, and membranous and particulate debris in the bile capillaries. The formation of megamitochondria and budding processes arising from them might be due to the inhibitory action of MTX upon nucleic acid formation in the mitochondria.

Ultrastructural Findings in the Bile Ducts in Liver Biopsies from 11 Psoriatics and 2 Controls. Bile ducts in liver biopsies from MTX-treated psoriatics do not seem to have been studied ultrastructurally before in a study of *Hopwood and Nyfors* [17] in 1977. They found in post-MTX liver biopsies from psoriatics atrophy of the bile duct-epithelial cells in the portal tracts, there being enlarged intercellular spaces with a consequent possibility of bile leakage. In the epithelial cells in the bile-duct epithelium there were *signs of damage,* viz. *damaged mitochondria,* a *hypertrophic and dilated Golgi apparatus,* and accumulation of membrane whorls in the cytoplasm. In places the bile-duct lumen was filled with membranous and particulate debris. Cilia were increased in number and showed varying number of central fibers. There was no necroses in the portal bile ducts and no abnormal bile-duct epithelium.

Studies of the Hepatotoxicity in Long-Term MTX-Treated Psoriatics

Zachariae et al. [62] in 1980 have published their liver biopsy findings in the largest group of long-term MTX-treated psoriatics. The purpose of the study was not to list the number of pathological findings in the liver biopsies.These questions seem to have been answered sufficiently in the above-mentioned studies [29, 30, 32, 58, 62, 63]. One of the purposes of the study of *Zachariae* et al. [62] in 1980 was to report the results of *serial liver biopsies in patients with established MTX-induced cirrhosis.* Data on serial liver biopsies from such 14 patients, of which 11 continued to receive MTX due to very severe psoriasis, seem to indicate that MTX induced liver cirrhosis is *not of a very aggressive nature. Hendel and Nyfors* [14] in 1985 came to similar results in their latest study where they did not find much progression of fibrosis in serial liver biopsies from long-term treated psoriatics. In 1982 *Ashton* et al. [3] published a study of 38 MTX-treated psoriatics with pre- and post-MTX liver biopsies: 7 patients (18%) developed fibrosis and 2 patients (5%) cirrhosis after an average cumulative MTX dose of 1,928 mg.

Factors Influencing Liver Histology

Alcohol Intake during MTX Intake

Long-lasting alcohol intake often produces liver-cell damage (degeneration + necrosis) and eventually cirrhosis [53]. Other workers [39, 56] have found alcohol intake to increase the tendency to fibrosis and cirrhosis in patients treated with MTX. It was natural, therefore, to investigate the significance of alcohol intake during MTX therapy of psoriasis. *Weinstein* et al. [58] in 1973 examined the relationship of alcohol intake and pathological findings in the liver in 204 pre-MTX patients. Increasing alcohol intake in these patients was significantly associated ($p < 0.035$) with the presence of fatty changes. In 338 post-MTX patients increasing alcohol intake significantly correlated with the presence of periportal inflammation, fibrosis, and cirrhosis. *Reese* et al. [43] in 1974 found a significant higher incidence of fatty change in patients who took alcohol during MTX medication ($p < 0.001$). In 92 patients with 1 post-MTX liver biopsy as well as in 68 having serial post-MTX biopsies, *Nyfors* [32] in 1977 found a significantly higher alcohol intake ($p < 0,05$) among the patients with cirrhosis than among those showing a normal histology of the latest liver biopsy. The pathogenetic role of alcohol in cirrhosis is further indicated by the findings in 2 patients who showed, apart from alcoholic hepatitis in several biopsies, also the co-existence of alcoholic hepatitis and micronodular cirrhosis in the latest biopsies [34]. *Zachariae* et al. [62] in 1980 found cirrhosis developing during MTX therapy in 19 patients, of which 3 had previously had a heavy alcohol consumption with an intake of more than 4 drinks (40 g of alcohol) a day. On the other hand, *Zachariae* et al. [62] in the same study found an improvement of the histology in a number of the latest liver biopsies in 328 psoriatics. The authors did not regard the mentioned improvement to be accidental, but to reflect a better following of their recommendations concerning alcohol. *Asthon* et al. [3] in 1982 found that their material of patients with fibrosis contained a high proportion of 'heavy drinkers' compared to those of not developing fibrosis. This finding supported the relation to high alcohol intake already noted by *Weinstein* et al. [58] in 1973, *Tobias and Auerbach* [56] in 1973, *Zachariae* et al. [61] in 1975, *Nyfors and Poulsen* [29] in 1976 and *Nyfors and Poulsen* [30] in 1976. In 1984, *Sørensen* et al. [54] published a long-term liver biopsy follow-up on 258 alcohol abusing men. These men, who were free from cirrhosis on primary liver biopsy, were followed for 10–13 years during which cirrhosis developed in 38, corresponding to a rate of 2% per year. The likelihood of cirrhosis developing proved to be independent of duration of abuse and of daily alcohol consumption before the primary biopsy. This indicates that the effect of alcohol abuse is not cumulative over time, but rather establishes conditions for the development of cirrhosis. The rate of cirrhosis increased stepwise with the degree of steatosis in the primary biopsy, and in those with alcoholic hepatitis was 9 times higher than in those with no steatosis. This finding, together with the results on alcohol abuse, indicates that steatosis and alcoholic hepatitis, despite their reversibility, are causally associated with cirrhosis rather than epiphenomena of alcohol abuse. In the alcohol abusing man, a liver biopsy provides more information than alcoholic history about the likelihood of future cirrhosis [54].

Obesity

Weinstein et al. [58] in 1973 found obesity significantly associated with fatty change in both pre-MTX and post-MTX biopsies. Among 14 patients exhibiting cirrhosis in the latest liver biopsy *Nyfors* [32] in 1977 (patient group B) found more overweight persons than among the 9 patients with normal histology at the latest liver biopsy ($p = 0.0033$).

Medicine

Hepatotoxic Medicine. Martin [24] in 1971 set down a long list of medicine that may influence liver function tests and thus be hepatotoxic.

Drug Interactions. The following drugs displace MTX from its plasma binding: *salicylates,* barbiturates, certain sulphonamides, *p*-aminobenzoic acid, and diphenylhydantoin [10, 24]. *Taylor and Halprin* [55] in 1977 found a mean MTX plasma albumin binding of 61% in 4 MTX-treated persons and 37% in the same persons during simultaneous administration of salicylate and MTX. The displacement of MTX from its plasma albumin binding entails a higher concentration of free MTX, which is rapidly redistributed into the total body water and is available for acting upon other tissues and for excretion. A higher level in plasma of free MTX may increase the risk of a toxic effect upon the organs. *Mandel* [23] in 1976 has reported 6 cases of pancytopenia – 2 with a fatal outcome – among 105 cancer patients who had been treated concomitantly with MTX and salicylates. Griseofulvin inhibits MTX via enzyme induction [24]. A finding of potential clinical importance is that other compounds, utilizing the same renal tubular transport mechanism as MTX, e.g. salicylate and weak organic acids such as *p*-aminohippurate, will diminish tubular transport of MTX [22]. The effect of MTX may be reduced in the event of concurrent administration of trimethoprim, which is also a folic acid antagonist. In the MTX guidelines, *Roenigk* et al. [48] have given the following list of drug interactions:

(A) Increased toxicity potential: (1) salicylate; (2) sulfonamides; (3) barbiturate; (4) *p*-aminobenzoic acid; (5) diphenylhydantoin; (6) probenecid (contraindicated); (7) phenylbutazone (contraindicated); (8) oral contraceptives; (9) colchicine, and (10) weak organic acids.

(B) Reduced toxicity potential: (1) tetracycline; (2) trimethoprim; (3) cephalothin; (4) hydrocortisone, and (5) griseofulvin.

In doses of MTX used for psoriasis, the altered toxic reactions are rarely clinically significant. The concomitant use of corticosteroids and MTX increases the risk of serious complications [48].

Cumulative Dose of MTX

The side effects of MTX appear to increase with the total dose to which the body is exposed. *Dahl* et al. [8] in 1971 found that the severity of liver damage in psoriatics was related to the duration of MTX treatment. *Almeyda* et al. [1] in 1972, in their material of post-MTX liver biopsies from 42 patients with psoriasis (one biopsy from each), found a significantly higher dose of MTX in patients with cirrhosis or fibrosis than among the other patients ($p = 0.05$). They found 3 cases of cirrhosis and 12 of fibrosis. *Nyfors* [32] in 1977 (patients group B), in 68 psoriatics with serial post-MTX biopsies, found a high percentage of cirrhosis (21%) as well as of fibrosis (24%). In this material the treatment period averaged 5½ years (and the total dose of MTX averaged about 4,000 mg). As already mentioned, similar, high incidences of fibrosis and cirrhosis were found by *Zachariae* [61] in 1975 (11% cirrhosis and 22% fibrosis). *Ashton* et al. [3] in 1982 found no statistically significant difference in the cumulative MTX dose between their patients with cirrhosis and without fibrosis/cirrhosis in their study of 38 patients with a pre- and post-MTX liver biopsy (after an average dose of 1,928 mg MTX).

Frequency of Administration

The accumulation of MTX depends, among other things, upon the interval at which the doses are administered. Since more than 4 half-times are required for complete exponential elimination, repeated administration of any drug at intervals shorter than this results in drug accumulation [13]. It has to be anticipated, therefore, that daily administration would entail a greater accumulation of the drug in the liver, and thereby more severe hepatotoxicity, than would weekly administration. This was confirmed in 1972 by *Dahl* et al. [9] who found more patients ($p = 0.05$) with hepatic fibrosis among 22 psoriatics treated with daily oral doses than among 14 psoriatics treated with weekly oral doses of MTX. Cirrhosis was present in 6 (14%) and hepatic fibrosis of varying severity in 11 (25%). The duration of MTX therapy varied form 3 to 74 months. *Podurgiel* et al. [40] in 1973 in a series of 33 psoriatics treated with oral MTX, demonstrated that cirrhosis and fibrosis were significantly less common if the drug was administered less than 12 times a month ($p < 0.01$). Cirrhosis was demonstrated in 12% and fibrosis in another 12% of the patients. The mean cumulative dose of MTX was 2,700 mg taken over a mean of 3 years. *Roenigk* et al. [47] in 1977, in 34 patients, found that fibrosis was more common after administration of oral MTX to psoriatics in 3 weekly doses (at intervals of 12 h) than in one weekly dose. Fibrosis was seen in 26% of post-MTX liver biopsies. The follow-up period was 2 years.

Administration of MTX in a single weekly oral dose, introduced by *Roenigk* et al. [45] in 1965 and by *Callaway* et al. [4] in 1966, seems to be the most favorable manner of therapy from the point of view of hepatotoxicity as well as of clinical experience. However, if the interval is further increased, the effect of MTX must be expected to decrease in several patients.

Dosage Schedules

These have been discussed in my first review [37]. *Weinstein* et al. [58] in 1973 gave substantial data for the fact that the daily MTX schedule was very hepatotoxic, and *Roenigk* et al. [47] in 1977 as well as *Zachariae* [63] in 1984 characterized the divided weekly dosage schedule as more hepatotoxic than the single weekly MTX dosage schedule.

Age

Roenigk et al. [46] in 1971 found that patients in whom abnormalities developed as shown on liver biopsies after MTX therapy tended to be an average of 15 years older than patients with normal results of liver biopsies. *Weinstein* et al. [58] in 1973 found that increasing age in pre- and post-MTX groups was not associated with the presence of fibrosis or cirrhosis. In 92 patients having 1 post-MTX liver biopsy, *Nyfors* [32] in 1977 (patients group A) found a significantly more advanced age among 7 patients with cirrhosis or fibrosis than among 13 patients with normal liver histology ($p < 0.01$).

Genetics

Besides the above-mentioned factors, other factors – at present known or unknown – as well as genetic factors may be of importance for the development of liver pathology [52]. *Nyfors and Sveigaard* [31] in 1976 looked for a possible relation between HL-A antigens and liver histology, but found no statistically significant association. *Zachariae* et al. [62] in 1980 found the antigen combination HLA-A1, B8 to be increased in patients with cirrhosis. A larger american study [yet unpubl.] has not been able to confirm the results of *Zachariae* et al. [62].

Control of Long-Term MTX-Treated Psoriatics

Liver Function Tests

Standard liver function tests are of little value in predicting the degree of liver damage as confirmed by several authors [12, 20, 25, 27, 29, 30, 32, 58, 61–63]. Other assessments of liver functions such as fasting bile salt [20] and galactose tolerance test [21] have been disappointing. A more positive attitude to extended liver function tests has been shown by *Rybach* et al. [49] in 1982 and by *Sørensen* et al. [54] in 1984.

Liver Ultrasound Scans

Neild et al. [28] in 1983, studying 18 MTX-treated psoriatics with liver ultrascan, concluded that patients with a normal ultrasound scan or moderate degrees of fatty change do not need liver biopsy. *Miller* et al. [25] in 1983 came to similar conclusions in a similar study of 32 psoriatics. *Geronemus* et al. [12] in 1982, in a study of 24 long-term MTX-treated psoriatics with technetium liver scan (isotopic scan) and liver biopsies, found that hepatotoxic reactions can not be reliably evaluated by this method.

Liver Biopsy

This seems to date, 1985, to be the only reliable mean of evaluating MTX-induced morphological liver changes. The morbidity from percutaneous liver biopsies ranges from 0 to 10% [12]. In patients with impaired coagulation, liver biopsy can be performed with plugging of needle trace [44].

Indices

Nyfors and Poulsen [29] in 1976 set up a *risk index I* for the probability of occurrence of pathological liver histology: twice height (cm) minus weight (kg) minus age (years) minus 50 (in case of daily alcohol intake) and minus 50 (in case of elevated SGOT). Fibrosis and cirrhosis is to be expected in patients with a low score in this risk index.

Nyfors and Poulsen [30] in 1976 set up *risk index II* comprising the 4 factors from *risk index I* + chief histological diagnosis of the pre-MTX liver biopsy (cirrhosis/possible cirrhosis/fibrosis 4 points, mixed changes 3 points, nonspecific reactive hepatitis 2 points, fatty change 1 point, and normal zero points). Index I should be used in studies of psoriatics in MTX therapy without a pre-MTX liver biopsy, while risk index II should be used in similar studies with a pre-MTX liver biopsy.

Zachariae et al. [62] in 1980 put up a 'cumulative cirrhosis index' composed of combined gradings of fibrosis, assessment of membrana limitans, fibrous destruction and regeneration to follow the development in more severe liver pathology in serial liver biopsies from the same patient.

Hendel and Nyfors [14] in 1985 have set up an index III with pharmacokinetic variables, among others. It seems as if the number of control liver biopsies can be cut down to half by using this index.

Prevention of Pathological Liver Changes during MTX-Therapy

Modesty in the use of alcohol seems to be very important as documented earlier in this paper. Personally, I would advise abstention from alcohol, allow maximum 5 drinks a month and deny initiate of

MTX therapy in patients who wanted or in terms of profession had to drink more than that amount of alcohol per month. Patients previously treated with arsenite are a high risk group [60]. Furthermore, the uttermost care in avoiding other hepatotoxic drugs should be shown by the patient's dermatologist, general practitioner, rheumatologist, etc. HL-A typing could be performed to see if the patient was A1, B8 type and thereby was more likely to develop cirrhosis.

Patients with lowered renal function belong to a risk group [48, 62]. Creatinine clearance is the most sensitive test of renal function [48].

Interpretation of Liver Biopsies

Interpretation of liver biopsies should be done by a pathologist being interested in liver histology [42, 50]. In patient with cirrhosis or severe to moderate fibrosis in their pre-MTX liver biopsy MTX treatment should not be started. In case of piecemeal necrosis the patient should only be given an additional 1–2 g of MTX before a new liver biopsy [35]. In case of piecemeal necrosis + moderate to severe fibrosis MTX therapy should be stopped [34]. In the 1980s there has developed an *acceptance of a broader and broader range of liver pathology compatible with continuation of MTX therapy. Zachariae* et al. [62] in 1980 have continued MTX treatment in psoriatics with cirrhosis and found the above-mentioned 'cumulative cirrhosis index' to decrease. The continuation seems to be justified in an interpretation of newer studies [14, 62] indicating that MTX is *not as progressively hepatotoxic as understood earlier* [29, 30, 32, 61]. The decrease of MTX hepatotoxicity in psoriasis seems to be due to a better selection of candidates for MTX therapy, omitting patients with a daily alcohol intake, previous arsenite intake, decreased renal function, etc.

Who Should Not Be Given MTX Therapy for Psoriasis?

Patients with daily intake of alcohol, salicylates, decreased renal function besides pregnancy and gastrointestinal ulcer.

Frequency of Control Liver Biopsy in MTX-Treated Psoriatics?

If one uses both of the above-mentioned risk indexes and ultrasound liver scans (when they are normal) the number of control liver biopsies can be reduced with about half compared to earlier apprehensions where a new liver biopsy was advised each time the dose passed a further 2 g of MTX [35]. In case of a liver-biopsy-proven liver pathology control liver biopsies should be performed after intake of a further 1.5–2 g of MTX [35, 62, 63]. The follow-up examinations in long-term MTX should be done in collaboration with pathologists, hepatologists, internists, rheumatologists etc. depending on the clinical problems during therapy.

Answers to Questions

(1) Cirrhosis is found in up to 3% of cases.

(2) No, at the age of 50 years, about 50% of pre-MTX liver biopsies will show pathological liver pathology: mainly fatty change and/or nonspecific reactive hepatitis.

(3) Up to 20–25% of fibrosis and up to about 20% of cirrhosis when treated for more than 5 years.

(4) Piecemeal necrosis, stellate periportal fibrous septa, fibrous septa from portal tracts to the adjacent portal tract or central vein, distortion of the lobular architecture, and eventually cirrhosis.

(5) It seems to have a nonaggressive course.

(6) Alcohol intake, concomitant intake of other hepatotoxic drugs, previous intake of arsenic, decreased renal function, obesity.

(7) Percutaneous liver biopsy preceded by ultrasound liver scans and use of the mentioned risk indices.

(8) Yes, the single weekly dose schedule is the least hepatotoxic.

(9) Omit patients with daily alcohol intake, renal dysfunction, previous arsenic intake, daily intakes of salicylates.

(10) Patients with cirrhosis, severe or moderate fibrosis, alcoholic hepatitis should not be given MTX treatment.

(11) By a team work between the different relevant specialities using strict exclusion criteria and with control of the liver status using with short intervals liver function tests and with longer intervals ultrasound scans, risk indexes and eventually percutaneous liver biopsy.

Conclusions

Hepatoxicity in terms of fibrosis after MTX therapy to patients with leukemia has been known since 1955. Cirrhosis was diagnosed in a post-MTX liver biopsy in a few psoriatics in the late 1960s. In the 1970s series of studies with a pre- and one or more post-MTX liver biopsies from the same MTX-treated psoriatic were published. Psoriatics seem to develop cirrhosis and fibrosis in up to 20% and 25% of cases, respectively, after a treatment period of 5–6 years. In the cirrhotic patients the morphological findings indicated the following sequence in the development of liver changes: first multiple, small piecemeal necroses with destruction of the limiting plate, followed by stellate periportal fibrosis. Later there developed portal-portal and portal-vein bridging, fibrosis with gradual alteration of the lobular architecture, leading ultimately to cirrhosis. Influental factors were shown to be alcohol intake, impairment of renal function, previous treatment with potassium arsenite, obesity, and age. In the 1980s the course of liver damage appeared to be nonaggressive. Newer studies seem to disclose less liver damage than previous studies. This may be due to a better following of our recommendations concerning alcohol as well as intensification of inclusion and exclusion criterea.

In the future, better selection of patients should take place with omission of patients with a daily alcohol intake. It should be possible to decrease the number of control liver biopsies (previously 1 per additional 2 g of MTX intake) by using ultrasonic liver scans and the described risk indexes. Perhaps HL-A typing can be used to find patients with a higher risk for developing cirrhosis.

References

1 Almeyda, J.; Bernardo, D.; Baker, H.; Levene, G.M.; Landels, J.W.: Structural and functional abnormalities of the liver in psoriasis before and during methotrexate therapy. Br. J. Derm. *87:* 623–631 (1972).

2 Anthony, P.P.; Ishak, K.G.; Nayak, N.C.; Poulsen, H.E.; Scheuer, P.J.; Sobin, L.H.: The morphology of cirrhosis. Definition, nomenclature, and classification. Bull. Wld Hlth Org. *55:* 521–540 (1977).

3 Ashton, R.E.; Millward-Sadler, G.H.; White, J.E.: Complications in Methotrexate treatment of psoriasis with particular reference to liver fibrosis. J. invest. Derm. *79:* 229–232 (1982).

4 Callaway, J.L.; McAfee, W.C.; Finlayson, G.R.: Management of psoriasis using methotrexate orally in a single weekly dose. South. med. J. *59:* 424–426 (1966).

5 Christoffersen, P.: Mallory legemer. Morfologi, forekomst og betydning (Fadl's Forlag, København 1976).

6 Coe, R.O.; Bull, F.E.: Cirrhosis associated with methotrexate treatment of psoriasis. J. Am. med. Ass. *206:* 1515–1520 (1968).

7 Colsky, J.; Greenspan, E.M.; Warren, T.N.: Hepatic fibrosis in children with acute leukemia after therapy with folic acid antagonists. Archs Path. *59:* 198–206 (1955).

8 Dahl, M.G.C.; Gregory, M.M.; Scheuer, P.J.: Liver damage due to Methotrexate in patients with psoriasis. Br. J. Med. *i:* 625–630 (1971).

9 Dahl, M.G.C.; Gregory, M.M.; Scheuer, P.J.: Methotrexate hepatotoxicity in psoriasis – comparison of different dose regimens. Br. J. Med. *i:* 654–656 (1972).

10 Dixon, R.L.; Henderson, E.S.; Rall, D.P.: Plasma protein binding of methotrexate and its displacement by various drugs. Fed. Proc. *24:* 454 (1965).

11 Epstein, E.H.; Croft, J.D.: Cirrhosis following methotrexate administration for psoriasis. Archs Derm. *100:* 531–534 (1969).

12 Geronemus, R.G.; Auerbach, R.; Tobias, H.: Liver biopsies vs, liver scans in methotrexate-treated patients with psoriasis. Archs Derm. *118:* 649–651 (1982).

13 Goodman, L.S.; Gilman, A.: The pharmacological basis of therapeutics; 5th ed. (MacMillan, New York 1975).

14 Hendel, J.; Poulsen, H.; Nyfors, B.; Nyfors, A.: Changes in liver histology during methotrexate therapy of psoriasis correlated to the concentration of methotrexate and folate in erythrocytes. Acta pharmac. tox. *56:* 321–326 (1985).

15 Hilden, M.; Christoffersen, P.; Juhl, E.; Dalgaard, J.B.: Liver histology in a 'normal' population – examination of 503 consecutive traffic casualities. Scand. J. Gastoent. *12:* 593–597 (1977).

16 Hopwood, D.; Nyfors, A.: Effect of methotrexate therapy in psoriatics on the Ito cells in liver biopsies, assessed by point counting. J. clin. Path. *29:* 698–703 (1976).

17 Hopwood, D.; Nyfors, A.: Liver ultrastructure in psoriatics related to Methotrexate therapy. II. Findings in bile ducts from 11 methotrexate-treated psoriatics and 2 controls. Acta pathol. microbiol. scand., A, Pathol. *85:* 801–811 (1977).

18 Horvath, E.; Kovacs, K.; Ross, R.C.: Liver ultrastructure in methotrexate treatment for psoriasis. Archs Derm. *108:* 427–428 (1973).

19 Hutter, R.V.P.; Shipkey, F.H.; Tan, C.T.C.; Murphy, M.L.; Chowdhury, M.: Hepatic fibrosis in children with acute leukemia. Cancer *13:* 288–307 (1960).

20 Lawrence, C.M.; Strange, R.C.; Summerly, R.; Scriven, A.J.; Elmahallawy, M.; Wood, A.; Fletcher, P.J.; Beckett, G.J.: Assessment of liver function using bile salt concentrations in psoriasis prior to and during methotrexate therapy. Clinica chim. acta *129:* 341–351 (1983).

21 Lenler-Petersen, P.; Søgaard, H.; Thestrup-Pedersen, K.; Zachariae, H.: Galactose tolerance test and methotrexate-induced liver fibrosis and cirrhosis in patients with psoriasis. Acta derm.-vener., Stockh. *62:* 448–449 (1982).

22 Liegler, D.G.; Henderson, E.S.; Hahn, M.A.; Oliverio, V.T.: The effect of organic acids on renal clearance of methotrexate in man. Clin. Pharm. Ther. *10:* 849–857 (1969).

23 Mandel, M.A.: The synergistic effect of salicylates on methotrexate toxicity. Plast. reconstr. Surg. *57:* 733–737 (1976).

24 Martin, E.W.: Hazards of medication. A manual on drug interactions, incompatibilities, contraindications and adverse effects, pp. 321, 678 (Lippincott, Philadelphia 1971).

25 Miller, J.A.; Dodds, H.J.; Lees, W.R.; Anderson, P.N.; Levene, G.M.; Munro, D.D.; Kirby, J.D.: A comparison of ultrasonography and liver biopsy in the assessment of methotrexate-induced hepatotoxicity in patients with psoriasis. Br. J. Derm., suppl. 24, p. 24 (1983).

26 Millward-Sadler, G.H.; Ryan, T.J.: Methotrexate induced liver disease in psoriasis. Br. J. Derm. *90:* 661–667 (1974).

27 Muller, S.A.; Farrow, G.M.; Martalock, D.L.: Cirrhosis caused by methotrexate in the treatment of psoriasis. Archs Derm. *100:* 523–530 (1969).

28 Neild, V.S.; Joseph, A.E.A., Marsden, R.A.; Fawcett, H.A.; Uff, J.S.: Liver ultrasound – a safe screening test for methotrexate patients. Br. J. Derm., suppl. 24, pp. 24–25 (1983).

29 Nyfors, A.; Poulsen, H.E.: Liver biopsies from psoriatics related to methotrexate therapy. I. Findings in 123 consecutive non-methotrexate treated patients. Acta pathol. microbiol. scand., A, Pathol. *84:* 253–261 (1976).

30 Nyfors, A.; Poulsen, H.E.: Liver biopsies from psoriatics related to methotrexate therapy. II. Findings before and after methotrexate therapy in 88 patients. A blind study. Acta pathol. microbiol. scand., A., Pathol. *84:* 262–270 (1976).

31 Nyfors, A.; Svejgaard, A.: The relation of HL-A antigens to liver histology in methotrexate-treated psoriatics. Acta derm.-vener., Stockh. *56:* 235–238 (1976).

32 Nyfors, A.: Liver biopsies from psoriatics related to methotrexate therapy. III. Findings in post-methotrexate liver biopsies from 160 patients. Acta pathol. microbiol. scand., A, Pathol. *85:* 511–518 (1977).

33 Nyfors, A.; Hopwood, D.: Liver ultrastructure in psoriatics related to methotrexate therapy. I. A prospective study of findings in hepatocytes from 24 patients before and after methotrexate treatment. Acta pathol. microbiol. scand., A, Pathol. *85:* 787–800 (1977).

34 Nyfors, A.; Poulsen, H.E.: Morphogenesis of fibrosis and cirrhosis in methotrexate-treated patients with psoriasis. Am. J. surg. Path. *1:* 235–243 (1977).

35 Nyfors, A.: Benefits and adverse drug experiences during long-term methotrexate treatment of 248 psoriatics. Dan. med. Bull. *25:* 208–211 (1978).

36 Nyfors, A.: Methotrexate therapy of psoriasis. Effects and side effects with particular reference to hepatic changes. A survey. Dan. med. Bull. *27:* 74–96 (1980).

37 Nyfors, A.: Methotrexate therapy in psoriasis and psoriatic arthritis. A review. Rheumatology, vol. 9, pp. 60–87 (Karger, Basel 1986).

38 O'Rourke, R.A.; Eckert, G.E.: Methotrexate-induced hepatic injury in an adult. Archs intern. Med. *113:* 191–194 (1964).

39 Palmer, H.M.: Hepatotoxicity of Methotrexate in the treatment of psoriasis. Practitioner *211:* 324–328 (1973).

40 Podurgiel, B.J.; McGill, D.B.; Ludwig, J.; Taylor, W.F.; Muller, S.A.: Liver injury associated with methotrexate therapy for psoriasis. Mayo Clin. Proc. *48:* 787–792 (1973).

41 Poulsen, H.E.: Histological features of acute viral hepatitis. Ann. clin. Res. *8:* 139–150 (1976).

42 Poulsen, H.E.; Christoffersen, P.: Atlas of liver biopsies (Munksgaard, Copenhagen 1979).

43 Reese, L.T.; Grisham, J.W.; Aach, R.D.; Eisen, A.Z.: Effects of methotrexate on the liver in psoriasis. J. invest. Derm. *62:* 597–602 (1974).

44 Riley, S.A.; Ellis, W.R.; Irving, H.C.; Lintott, D.J.; Axon, A.T.R.; Losowsky, M.S.: Percutaneous liver biopsy with plugging of needle track. A safe method for use in patients with impaired coagulation. Lancet *ii:* 436 (1984).

45 Roenigk, H.H., Jr; Haserick, J.R.; Curtis, G.H.: Methotrexate for psoriasis. A preliminary report. Cleve. Clin. Q. *32:* 211–215 (1965).

46 Roenigk, H.H., Jr.; Bergfeld, W.F.; Jacques, R.S.; Owens, F.J.; Hawk, W.A.: Hepatotoxicity of methotrexate in the treatment of psoriasis. Archs Derm. *103:* 250–261 (1971).

47 Roenigk, H.H., Jr.; Auerbach, R.; Bergfeld, W.F.; Frost, P.; Maibach, H.I.; Tindall, J.; Weinstein, G.D.: A co-operative, prospective study of the effects of chemotherapy of psoriasis on liver biopsies; in Farber, Cox, Psoriasis. Proc. 2nd Int. Symp., pp. 243–248 (Yorke Medical Books, New York 1977).

48 Roenigk, H.H., Jr.; Auerbach, R.; Maibach, H.I.; Weinstein, G.D.: Methotrexate guidelines – revised. J. Am. Acad. Derm. *6:* 145–155 (1982).

49 Ryback, R.S.; Eckardt, M.J.; Felsher, B.; Rawlings, R.R.: Biochemical and hematologic correlates of alcoholism and liver disease. J. Am. med. Ass. *248:* 2261–2265 (1982).

50 Scheuer, P.J.: Liver biopsy interpretation; 2nd. ed. (Bailliere & Tindall, London 1973).

51 Scheuer, P.J.; Thaler, H.: Acute and chronic hepatitis revisited. Review by an international group. Lancet *ii:* 914–919 (1977).

52 Sherlock, S.: Diseases of the liver and biliary system; 6th ed. (Blackwell, Oxford 1981).

53 Shiff, L.: Diseases of the liver; 4. ed. (Lippincott, Philadelphia 1975).

54 Sørensen, T.I.A.; Orholm, M.; Bentsen, K.D.; Høybye, G.; Eghøje, K.; Christoffersen, P.: Prospective evaluation of alcoholic abuse and alcoholic liver injury in men as predictors of development of cirrhosis. Lancet *ii:* 241–244 (1984).

55 Taylor, J.R.; Halprin, K.M.: Effect of sodium salicylate and indomethacin on methotrexate serum albumin binding. Archs Derm. *113:* 588–591 (1977).

56 Tobias, H.; Auerbach, R.: Hepatotoxicity of long-term methotrexate therapy for psoriasis. Archs intern. Med. *132:* 391–396 (1973).

57 Warin, A.P.; Landells, J.W.; Levene, G.M.; Baker, H.: A prospective study of the effects of weekly oral methotrexate on liver biopsies. Findings in severe psoriasis. Br. J. Derm. *93:* 321–327 (1975).

58 Weinstein, G.; Roenigk, H.; Maibach, H.; Cosmides, J.: Psoriasis-liver-methotrexate interactions. Archs Derm. *108:* 36–42 (1973).

59 Zachariae, H.; Søgaard, H.: Liver biopsy in psoriasis. A controlled study. Dermatologica *146:* 149–155 (1973).

60 Zachariae, H.; Søgaard, H.; Nyfors, A.: Liver biopsy in psoriatics previously treated with potassium arsenite. A controlled study. Acta derm.-vener., Stockh. *54:* 235–236 (1974).

61 Zachariae, H.; Grunnet, E.; Søgaard, H.: Liver biopsy in methotrexate-treated psoriatics – a re-evaluation. Acta derm.-vener., Stockh. *55:* 291–296 (1976).

62 Zachariae, H.; Kragballe, K.; Søgaard, H.: Methotrexate induced liver cirrhosis. Br. J. Derm. *102:* 407–412 (1980).

63 Zachariae, H.: Psoriasis and the liver; in Roenigk, Jr., Maibach, Psoriasis (Dekker, New York 1985).

Prof. Allan Nyfors, MD, Department of Dermatology,
Haukeland University Hospital, N–5016 Bergen (Norway)

Rheumatology, vol. 9, pp. 213–221 (Karger, Basel 1986)

Liver Biopsy Findings after Prolonged Methotrexate Therapy for Rheumatoid Arthritis

Allen H. Mackenzie

Clinical Section, Department of Rheumatic and Immunologic Disease, Cleveland Clinic Foundation, Cleveland, Ohio, USA

Methotrexate (MTX) has proven empirically useful in suppressing the inflammation of severe, progressive, rheumatoid disease [1]. Rheumatoid arthritis (RA) is considerably suppressed within the first few weeks of treatment, and tends to relapse promptly upon discontinuation, its mode of action remains speculative [2].

Since this treament had proven to be so useful, large numbers of patients with RA had received MTX between 1963 and 1973. Their disease had relapsed so consistently upon its discontinuation, that a need for the assessment of safety of chronically used MTX became necessary. Reports from the dermatology literature employing MTX for psoriasis vulgaris had indicated a possibility of hepatic injuries including portal fibrosis, and even cirrhosis [3–6]. That literature had indicated that liver biopsy was necessary to detect these changes. A study was therefore undertaken in 1973 to perform liver biopsy in rheumatoid patients being treated with MTX to assess the long-term safety of this treatment [7, 8].

Methods and Data

The hazards of MTX therapy and of liver biopsy were discussed with patients having RA, who had consumed MTX for 2 or more years of continuous therapy. It was possible to obtain liver biopsy in 60 such patients on chronic MTX therapy, together with an additional 25 rheumatoid patients about to embark on MTX therapy, so that 60 patients on chronic therapy with MTX could be compared with 25 controls having comparably severe RA, but never having been treated with MTX.

Table I. Series comparison

	25 RA patients no MTX Rx		60 RA patients >24 months MTX Rx	
Age mean, years	50.8	(21–70)	52.5	(16–76)
RA duration, years	8.1	(1–26)	9.4	(3–31)
Female, n (%)	16	(64)	36	(64)
Male, n (%)	9	(36)	24	(40)
Latex +, n (%)	21	(84)	47	(78)
ANA +, n (%)	12	(48)	29	(54)[1]
Sjogren's, n (%)	5	(20)	16	(27)
Nodules, n (%)	8	(32)	24	(40)
Erosive, n (%)	23	(92)	52	(87)
Polyarticular and persistent	25	(100)	60	(100)[2]
Alcohol use: 0–minimal	17	(68)	41	(69)
Alcohol use: moderate–heavy	8	(32)	19	(31)

[1] Not done in 6 patients.
[2] Spontaneous remission previously in 3 patients.

MTX is used in combination with an antimalarial and a nonsteroidal drug. Antidepressants and corticosteroids are combined less frequently. Allergy to MTX has not been observed. Side effects have been kept quite low by dividing dosage, by keeping the weekly dosage low, and by giving supplemental folate. The safety record of this therapy has been gratifying with occasional mild stomatitis or macrocytosis, and rare elevations of transaminase being seen. These manifestations have been interpreted as folate depletion, and have tended to respond to intermittent folate therapy. No serious bone marrow depression or macrocytic anemia, no proctitis or hepatitis have been observed at the prescribed dosage. No alteration of menstrual pattern has been seen. It is my impression that there is a modest increase in the frequency of herpes virus infection during therapy. Other infections of the type seen in immunosuppressed patients do not occur with increased frequency. The overall dropout rate during MTX therapy has remained low, estimated at 6% of all patients.

Other medications in the study were comparable, including almost invariably, a salicylate or nonsteroidal antiinflammatory drug, an antimalarial compound, either chloroquine or hydroxychloroquine, in dosage up to, but not exceeding 4 mg/kg/day chloroquine, or 6.5 mg/kg/day hydroxychloroquine. Thirteen patients were on low-dose corticosteroid therapy never exceeding 7.5 mg prednisone per day. Between 1963 and 1968, no folic acid was given, but beginning in 1968, folic acid replacement was gradually added [1], so that by the middle of 1969, all patients were receiving folic acid, 10–15 mg/day for 1 week out of each 8-week therapy cycle, during which time MTX was omitted, and take MTX for 7 weeks out of each 8-week cycle.

Table II. Duration and quantity of MTX ingestion

	Number of patients	Mean age years	Mean duration RA, years	Mean months of MTX therapy (range)	Mean mg MTX ingested (range)
	18	54	7.4	28.6 (24–36)	905 (505–1,810)
	13	54	11.2	40.3 (36–48)	1,650 (1,128–2,071)
	11	52	8.5	53.3 (48–60)	2,023 (1,358–3,946)
	11	54	9.1	66.4 (60–72)	2,630 (1,239–3,939)
	7	46	10.9	77.8 (72–94)	3,740 (1,856–4,440)
Total mean		52.5	9.4	48.5	1,837
Total	60			2,903	110,263

In table I, comparison of the treated series with the control group reveals similar severity of rheumatoid disease with similar sex ratios, rheumatoid factor seropositivity, erosive change, persistence of synovitis, and alcohol consumption.

The treated group are about 3 years older than the control group, had a few more rheumatoid nodules, and a slight excess of Sjogren's syndrome features. The series of treated and control patients appear quite comparable. This is the severe end of the spectrum of destructive rheumatoid disease, poorly responsive to prior therapy.

Methotrexate Administration

In table II, a summary of MTX ingestion is presented showing the number of patients, age, duration of rheumatoid, and MTX therapy in months, and of mean cumulative MTX ingested in milligrams. No patient received less than 24 months of MTX therapy in this series; there were 18 patients who had taken MTX between 2 and 3 years, 13 between 3 and 4 years, 11 between 4 and 5 years, 11 between 5 and 6 years, and 7 more than 6 years. The mean duration of MTX therapy in this series was thus 48.5 months (range 24–94), and mean consumption of drug was 1,837 mg, ranging between 505 mg and 4.44 g. Mean weekly dose was 8.67 mg (range 2.5–12.5 mg).

Table III. Classification of liver biopsy findings [9]

Grade	Pathological findings	
I	normal	mild
	fatty infiltration	mild
	nuclear variability	mild
	portal inflammation	mild
II	fatty infiltration	moderate–severe
	nuclear variability	moderate–severe
	portal tract expansion	
	portal inflammation	moderate–severe
	focal necrosis	
III	fibrosis (septum formation)	
IV	cirrhosis	

Liver biopsies were obtained percutaneously by standard needle biopsy techniques using Menghini or Vim-Silverman needles. The specimens were fixed and stained using standard methods with hematoxin-eosin, as well as Masson trichome stains. Pathologists examined the biopsy material, in the knowledge that MTX was in question, but not knowing whether the biopsy was obtained prior to or after MTX therapy. Liver biopsies were classified according to the National Program for Dermatology Guidelines [9], in which liver findings are graded I–IV in degree of severity (table III).

The dermatologists have concluded that grade II liver changes, even fibrous expansion of the portal zone, is an indication for liver rebiopsy in the future, but not to discontinue MTX therapy. Grade III changes (fibrosis, formation of fibrous septa between liver lobules), however, are regarded as a contraindication against further MTX therapy.

In the present study, all of the liver biopsies in both the MTX and the control groups, fell within grades I and II, although one MTX-treated patient did have a minor expansion of the portal tract with fibrous tissue, no septum formation was detected.

In table IV, the summary of histopathology of the liver in all 85 rheumatoid patients, portal inflammation occurred in about 20% of severe rheumatoids, not different in the MTX and untreated groups. Fatty infiltration of the liver occurred in about 50% of the MTX-

Table IV. Summary of liver biopsy findings

	60 MTX Rx		25 controls	
	n	%	n	%
Normal	26	43	12	48
Anisonucleosis	13	22	3	12
Fatty infiltrates	30	50	11	44
No fat	30	50	14	56
Portal inflammation	11	18	5	20
No PI	48	80	21	80
Hepatic granuloma	1	1.7	0	
Portal fibrosis	1	1.7	0	
Cirrhosis	0		0	

treated group, and about 44% of the control patients. A single case of scattered, noncaseating, hepatic granulomata was observed, which had disappeared a year later upon rebiopsy. Expansion of the portal zone with fibrous tissue was observed in one MTX-treated patient, but did not reveal interlobular fibrous septum formation. This patient was a 56-year-old man with RA of 7 years duration, who consumed MTX for 57 months, for a cumulative total of 2.8 g. He had well-developed Sjogren's syndrome and very chronic peptic ulcer disease. On follow-up, this individual was not rebiopsied, but developed no evidence of progressive hepatic disease.

Fatty Liver

Fatty infiltrates in the liver did not appear to correlate with MTX therapy, although 6% more of the MTX treated group had steatosis than of the controls. The cumulative mean MTX dose was the same for all categories of none, minimal and mild (grade I), moderate and marked (grade II) fatty infiltration. Fatty infiltration occurred more frequently in older patients (table V) and in consumers of alcohol (table VI).

69% of drinkers had some degree of fat in the liver, while only 41% of teetotalers had fatty infiltrates. The presence of Sjogren's syndrome

Table V. Fatty liver in 60 RA patients MTX Rx

Pathology	Grade	Number of patients	Cumulative mean MTX mg	Mean duration of MTX Rx months	Mean duration of RA years	Mean age of patients years
None		30	1,835	48.1	8.5	49.3
Minimum-mild	I	25	1,840	49.1	10.6	55.4
Moderate-marked	II	5	1,844	46.6	8.8	57.6
Total		60	1,837	48.5	9.4	52.5

Table VI. Fatty liver biopsy; alcohol use by 85 RA patients

Pathology	MTX Rx group		Controls	
	n	%	n	%
Nondrinkers				
Lack fat	24	59	11	65
Have fat	17	41	6	35
Drinkers				
Lack fat	6	31	3	37
Have fat	13	69	5	63
Total	60		25	

Table VII. Portal inflammatory infiltrates (PI)

Pathology	Grade	MTX cumulative mean dose, mg	Mean duration of MTX Rx months	Number of patients	Mean duration of RA years	Mean age of patients years
None	I	1,690	45	48	9.7	52.4
Min PI	I	3,070	69	3	8.3	48
Mild PI	I	2,220	63.5	6	7.2	50.5
Marked PI	II	1,915	53	2	12	65
Fibrosis	III	2,838	57	1	7	56
Total		1,837	48.5	60	9.4	52.5

Table VIII. Anisonucleosis and anisocytosis

Number	MTX Rx					
	portal infiltration	%	mean %	fatty infiltration	%	mean %
24	males have PI	16		have fat	68	
36	females have PI	23	20	have fat	44	48
21	Sjogren's have PI	38		have fat	62	
Controls						
9	males have PI	22		have fat	55	
16	females have PI	19	20	have fat	37	44
5	Sjogren's have PI	20		have fat	60	

Table IX. Anisocytosis-anisonucleosis in liver biopsy (85 patients)

		n	%		n	%
60 patients MTX Rx	equal cells	47	78	ANISO	13	22
25 patients Controls	equal cells	22	88	ANISO	3	12

Alcohol consumption (one-third of each group) did not influence anisocytosis.

features, peptic ulceration and diabetes mellitus further amplified the amount of fatty infiltration.

Portal Infiltration

Infiltration of the portal zone with round cells was seen frequently. Although one might expect an increase of such biopsy evidence of cellular inflammation to precede fibrosis, no trend could be detected in this data to indicate that (table VII). The numbers of patients are too small to indicate whether alcohol use influences portal infiltration.

The data do not suggest that increased cumulative MTX dose is a factor increasing portal infiltration. Advancing age may be a factor in portal infiltrates. In addition, Sjogren's syndrome may be a factor increasing portal infiltrates, at least in the MTX-treated group (table VIII).

Unequal size of liver cells or liver nuclei was present in 12% of the control patients, and 22% of those treated with MTX. This is definitely a feature of folate deficiency, as might be expected from the clinical manifestations in folate-depleted states (table IX).

Discussion

In these 60 patients, the mean weekly dose of MTX was 8.67 mg (median dose 7.5 mg/week). This is roughly a third of the dosage at which the experience in treating psoriasis vulgaris with MTX was obtained. The results of this study appear to confirm the clinical impression that low-dose MTX therapy administered for between 2 and 8 years, used as suppressive therapy for chronic RA, has relatively minor harmful effects upon liver histopathology. A possibility that concurrent salicylate therapy might amplify toxicity could not be controlled in this study, but the potential interaction of salicylates with high-dose MTX therapy appears not to be clinically significant at the low doses used. The possibility that alternating co-therapy with folic acid may ameliorate MTX hepatotoxicity was uncontrolled in this study, and remains unproven.

Conclusions

Long-term (4 years mean), low dose (8.7 mg/week mean) MTX maintenance therapy for RA had a minor and clinically unimportant hepatotoxic effect in this patient series. A single case of portal fibrosis without septum formation was observed, but liver disease was nonprogressive during a follow-up period of 4 years. The majority of hepatic histopathology in this series appears to be related to the severity of RA, to advancing age, to concomitant Sjogren's syndrome, and to alcohol consumption. Since therapy with MTX appears to be reasonably safe during this study: (a) the efficacy of MTX in RA needs to be documented via controlled clinical study; (b) the effects of co-therapy with other agents, in particular the folate, require additional study; (c) still longer-term therapy needs to be studied in order to determine the magnitude and clinical significance of lower levels of toxicity, and (d) follow-up studies on those who do develop portal fibrosis are needed to determine whether this lesion may be progressive, and just how harmful it is.

Note: Methotrexate was not a popular therapy for RA in 1974. It has subsequently become more so. Studies on MTX used for RA are accumulating [10–12]. Therefore, this report was submitted in 1984 to provide more detail, as the information had been published only in abstract form [7, 8].

References

1 Mackenzie, A.H.; Scherbel, A.L.: Management of rheumatoid arthritis in the surgical patient. Orthop. Clin. N. Am. *2:* 277–299 (1971).

2 Gubner, R.; August, S.; Ginsburg, V.: Therapeutic suppression of tissue reactivity I and II. Am. J. med. Sci. *221:* 169–182 (1951).

3 Hersch, E.M.; Wong, V.G.; Henderson, E.S.; Freireich, E.J.: Hepatotoxic effects of methotrexate. Cancer *19:* 600–606 (1966).

4 Roenigk, H.H.; Bergfeld, W.F.; St. Jacques, R.; Owens, R.J.; Hawk, W.A.: Hepatotoxicity of methotrexate in the treatment of psoriasis. Archs Derm. *103:* 250–261 (1971).

5 Weinstein, G.; Roenigk, H.; Maibach, H.; Cosmides, J.; Halprin, K.; Millard, M.: Psoriasis-liver-methotrexate interactions. Archs Derm. *108:* 36–42 (1973).

6 Podurgiel, B.J.; McGill, D.B.; Ludwig, J.; Taylor, W.F.; Muller, S.A.: Liver injury associated with methotrexate therapy for psoriasis. Mayo Clin. Proc. *48:* 787–792 (1973).

7 Mackenzie, A.H.: Liver biopsy findings after methotrexate therapy for rheumatoid arthritis (Abstract). J. Rheumatol. *1:* suppl. 1, p. 73 (1974).

8 Mackenzie, A.H.: Liver biopsy findings after methotrexate therapy for rheumatoid arthritis (Abstract). Clin. Pharmacol. Ther. *17:* 239 (1975).

9 Roenigk, H.H.; Maibach, H.I.; Weinstein, G.D.: Guidelines, methotrexate therapy for psoriasis (Pamphlet). (National Program for Dermatology, 4 November 1972).

10 Wilke, W.S.; Calabrese, L.H.; Scherbel, A.L.: Methotrexate in the treatment of rheumatoid arthritis. Cleve. Clin. Q. *47:* 305–321 (1980).

11 Willkens, R.F.; Watson, M.A.: Methotrexate: a perspective of its use in the treatment of rheumatic diseases. J. Lab. clin. Med. *100:* 314–321 (1982).

12 Groff, G.D.; Shenberger, K.N.; Wilke, W.S.; Taylor, T.H.: Low dose oral methotrexate in rheumatoid arthritis. An uncontrolled trial and review of the literature. Semin. Arthritis Rheum. *12:* 333–347 (1983).

Allen H. Mackenzie, MD, Department of Rheumatic and Immunologic Disease, Cleveland Clinic Foundation, 9500 Euclid Avenue, Cleveland, OH 44106 (USA)

Rheumatology, vol. 9, pp. 222–234 (Karger, Basel 1986)

Liver Biopsy Findings during Methotrexate Therapy of Rheumatoid Arthritis

H. Frenzel[a], *T. Karger*[b], *R. Rau*[b]

[a] Pathologisches Institut der Universität, Düsseldorf, BRD; [b] Rheumaklinik, Evangelisches Fachkrankenhaus, Ratingen, BRD

Introduction

Since 1958, methotrexate (MTX) has been used in the treatment of severe psoriasis vulgaris [7] and encouraging results have been recorded in many clinical studies [2, 13, 21, 26, 28]. However, the application of MTX in low doses is not free of side effects preferably affecting the liver. In biopsies, fibrosis and cirrhosis of the liver could be shown in patients with psoriasis vulgaris in the course of MTX treatment [11, 14, 17, 19–21, 27, 32], but the frequency of those complicating alterations varied. *Dahl* et al. [4] found fibrosis of the liver in 27% and cirrhosis in 19% of their psoriatic patients who had received MTX in low dose by different regimen. From clinical studies it was evident that the frequency and degree of these side effects may be influenced by various other factors [5, 12–14, 21].

Low-dose application of MTX has also been proven to be useful in the treatment of rheumatoid arthritis (RA) [9, 10, 25, 29, 30]. Differing from MTX treatment of psoriatic patients, only minor side effects have been reported in patients with RA who had received MTX [6, 8]. However, there are only a few studies concerning side effects of MTX on the liver histology of patients with RA.

In the present study the side effects of MTX on the liver histology were evaluated comparing liver biopsies of patients with RA who had been treated with or without MTX. Because, in patients with RA, liver damage may have various causes, an attempt was made to define those

Table I. RA patients without MTX treatment

Number of patients	10 (4 M/6 F)
Age	55.7 ± 12.1 years (35–70)
Duration of RA	5.3 ± 4.1 years (1–15)
Activity of RA	mild 2/10, moderate 4/10, severe 4/10
Therapy	NSARD 1/10, NSARD + gold 6/10
	NSAR + d-PA 1/10, NSAR + gold + steroids 2/10

Table II. RA patients with MTX treatment

Number of patients	17 (2 M/15 F)
Age	56.7 ± 8.6 years (46–74)
Activity of RA	mild 35%, moderate 47%, severe 18%
Duration of MTX-application	27.9 ± 9.4 months (4–45)
Cumulative MTX dose	1,333.5 ± 610 mg (280–1,920)
Concomitant therapy	NSARD 24%, NSARD + gold 23%
	NSARD + steroids 35%
	NSARD + gold + steroids 18%
GPT, GOT elevation	41% (36–238 U/1)

alterations which can be related with certainty to MTX. Finally, the study should give information whether repeated liver biopsies in MTX-treated patients with RA are required.

Material and Methods

Liver biopsies of patients with RA of various duration and severity were semi-quantitatively evaluated.

RA Patients without MTX Treatment

This group includes 10 patients (4 men and 6 women) with a mean age of 55.7 ± 12.1 years. The mean duration of RA was 5.3 ± 4.1 years, the RA was classified as moderate or severe in 80% of the patients according to *Voit and Gamp* [1958]. All patients had received various therapeutic compounds over a different period of time, in 60% a combination therapy (NSARD + gold) was applied at the time the liver biopsy was taken. MTX was not given (table I).

RA Patients with MTX Treatment

This group includes 17 patients (2 men, 15 women) with a mean age of 56.7 ± 8.6 years. The RA was classified as moderate or severe in 65% of patients. All patients re-

Table III. Liver biopsy changes in RA patients and RA patients with MTX treatment; frequency (%) and severity (score) of morphological variables

Morphological variables	RA without MTX		RA with MTX	
	frequency	score	frequency	score
Nuclear variability*	80	0.8	94	1.2
Ballooning	80	0.8	76	1.06
Granular cytoplasm	60	0.6	94	1.1
Fatty change*	60	1.0	71	0.88
Liver cell necrosis*	40	0.4	35	0.35
Cholestasis	0	0	6	0.06
Peribiliar basophilia	10	0.1	35	0.35
Kupffer cell prolif.	100	1.4	100	1.6
PT infiltration*	80	1.0	82	0.8
PT fibrosis*	30	0.5	47	0.47
Lipofuscin	90	1.1	100	1.5
Pigment Kupffer cells	70	1.1	88	1.3
Pigment portal tracts	30	0.4	65	1.1
Iron	10	0.1	18	0.35
Limiting plate disruption	10	0.1	0	0
Total score/patient	9.5 ± 3.5		12.2 ± 3.1	
Special score/patient	3.7 ± 2.6		3.7 ± 1.3	

*Variables of special score.

ceived MTX in combination with NSARD, gold and/or steroids. The weekly dose was given in 3 single applications with 12-hour intervals between the doses. The mean duration of MTX therapy was 27.9 ± 9.4 months, the mean cumulative dose was 1,333.5 ± 610 mg. In 41% of the patients an elevation of GOT and GPT activity was observed during the MTX treatment with a maximum of 238 U/l. Patients with increased alcohol consumption were excluded from the study. In no case was liver biopsy taken because of a striking transaminase elevation (table II).

Liver Biopsy

Liver biopsies were obtained by the standard technique using a Menghini needle. The specimens were fixed in 4% formalin, and paraffin sections were stained using standard methods with hematoxylin and eosin, elastica-van Gieson, silver stain according Gomori, PAS, Ladewig for collagen fibres, Berliner blue for iron, and Shikata reaction for demonstration of virus B surface antigen.

15 histological variables (table III) were evaluated semiquantitatively with the liver findings being graded from 0 to 3 in degree of severity. Summing up the values for variables, a score was obtained for each patient.

Results

RA without MTX Treatment

In the biopsies of all 10 patients the typical lobular pattern of liver tissue was preserved. Without exception, a non specific reactive hepatitis with activated and proliferated Kupffer cells and portal tract infiltration of lymphocytes and monocytes was observed. A mild portal or periportal fibrosis could be recognized in the biopsies of 3 patients. Hepatocellular alterations consisted of ballooning of liver cells (8/10) or fine granulation (6/10) of their cytoplasm besides single coagulation necrosis, so-called Councilman bodies (4/10), in a peribiliar basophilia (10/10) indicating a disturbed bile excretion, and in a mild or moderate fatty change in 6 patients. 8 of 10 patients showed an increased variability of the nuclear size and polyploidized nuclei of liver cells in the pericentral region. Two kinds of pigment were deposited in the liver tissue of nearly all patients: lipofuscin in the pericentral hepatocytes and a granular yellow-brownish pigment in the Kupffer cells and macrophages of portal tracts (table III).

RA with MTX Treatment

None of the patients showed liver tissue devoid of any alteration. Liver cirrhosis could not be observed while a mild portal tract fibrosis was seen in 47% of the patients. The periportal limiting plate was intact, piecemeal necrosis was lacking. One patient showed a minimal cholestasis, another one a small granuloma. The liver biopsies of nearly all patients disclosed mesenchymal reactions (proliferation of Kupffer cells in 100% and inflammatory infiltration of portal tracts in 82% (fig. 1) and hepatocellular alterations (cellular swelling in 76%, increased variability of nuclear size in 94%, necrosis in 35%, fatty change in 74% (fig. 2) as seen in the group of patients without MTX treatment. The pigment deposits (fig. 3) corresponded qualitatively to those in the RA patients without MTX application but were more abundant (table III).

Comparison of Both Groups of Patients

Most of the evaluated morphological variables were found in both groups with the same frequency (table III). In addition, the semiquantitative score of each of these morphological variables is listed in table III. In the semiquantitative evaluation the degree of severity of

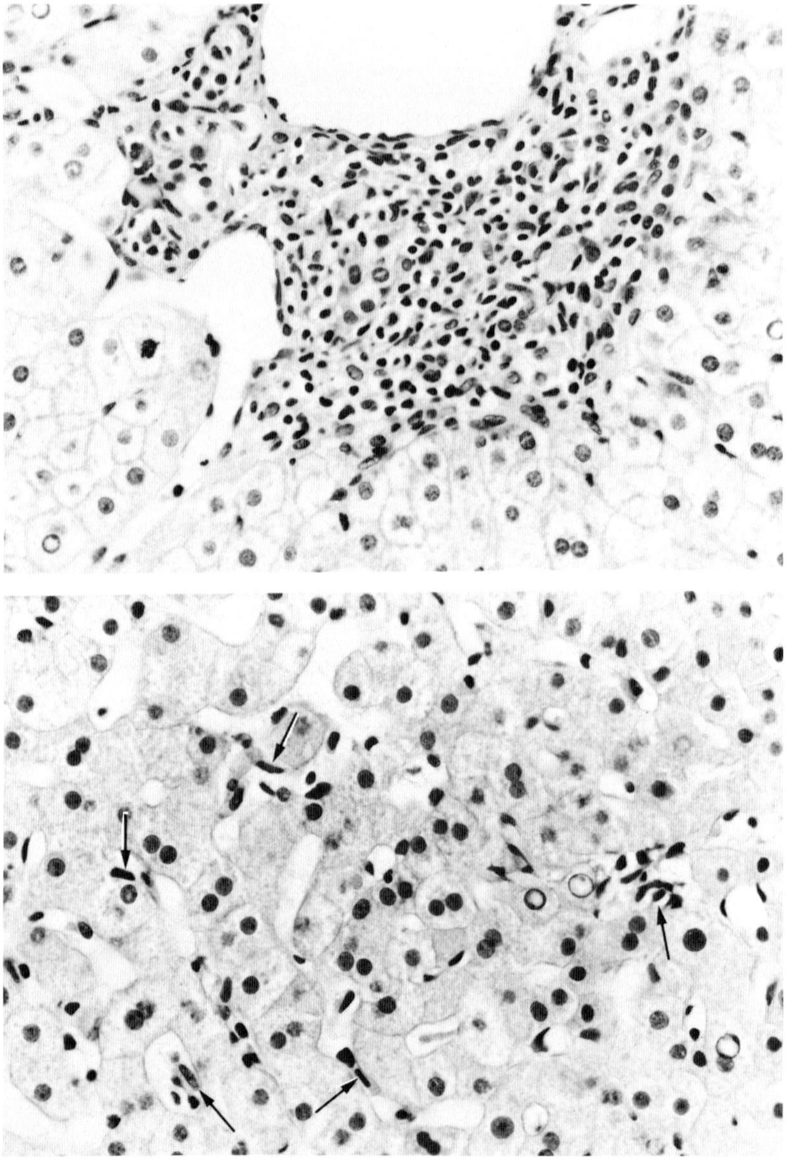

a

b

Fig. 1. Mesenchymal alterations of liver parenchyma during MTX treatment of RA revealed in liver biopsies. *a* Moderate portal tract infiltration of lymphocytes and a few plasma cells. The limiting plate is intact. HE. × 290. *b* Proliferated Kupffer cells in the sinusoids (arrows). HE × 290.

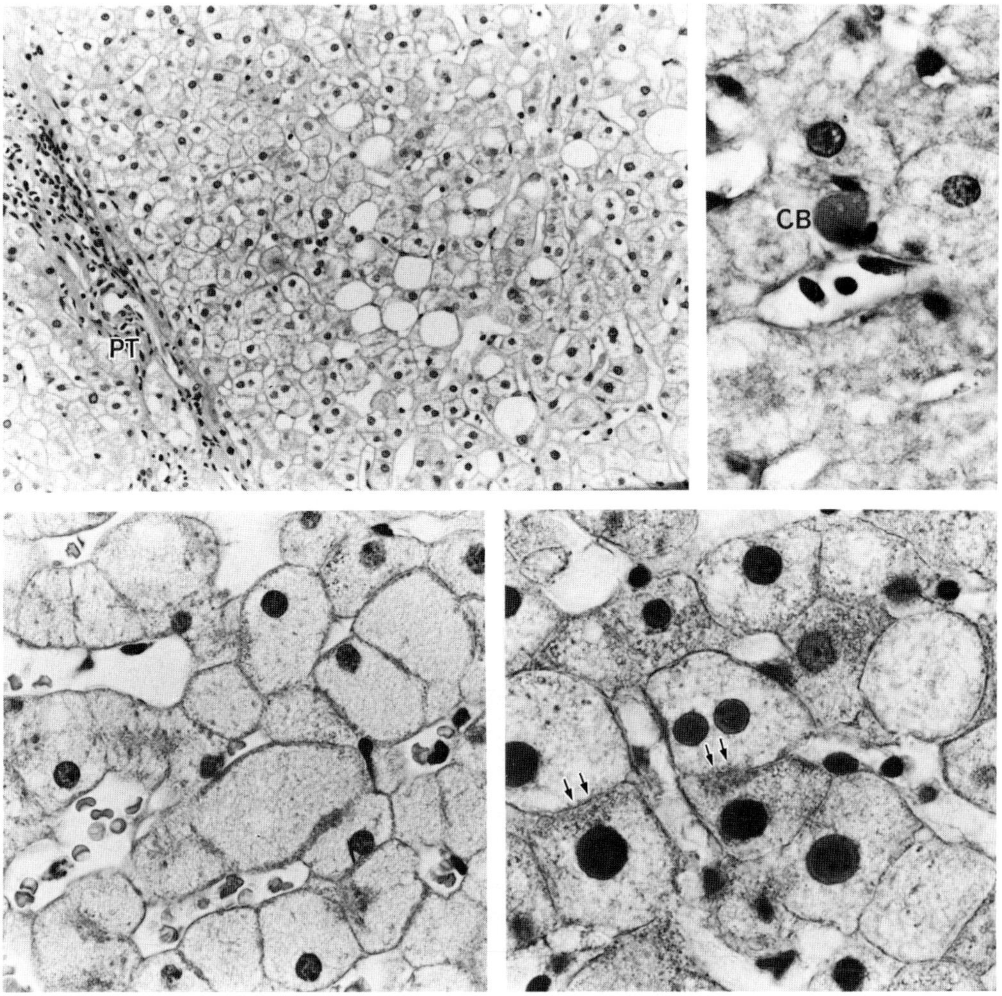

Fig. 2. Hepatocellular alterations during MTX treatment of RA revealed in liver biopsies. *a* Mild to moderate fatty change of hepatocytes in the pericentral lobular region. Slight lymphoid infiltration of a portal tract (PT). The liver cell nuclei show a mild variability of their size. HE × 140. *b* Acidophilic liver cell necrosis, so-called Councilman body (CB), surrounded by Kupffer cells. HE × 610. *c* Fine granulation of the cytoplasm of hepatocytes corresponding to a proliferation of endoplasmic reticulum. HE × 410. *d* Hepatocytes with swollen cytoplasm, increased variability of nuclear size, and peribiliary basophilia (arrows). HE × 540.

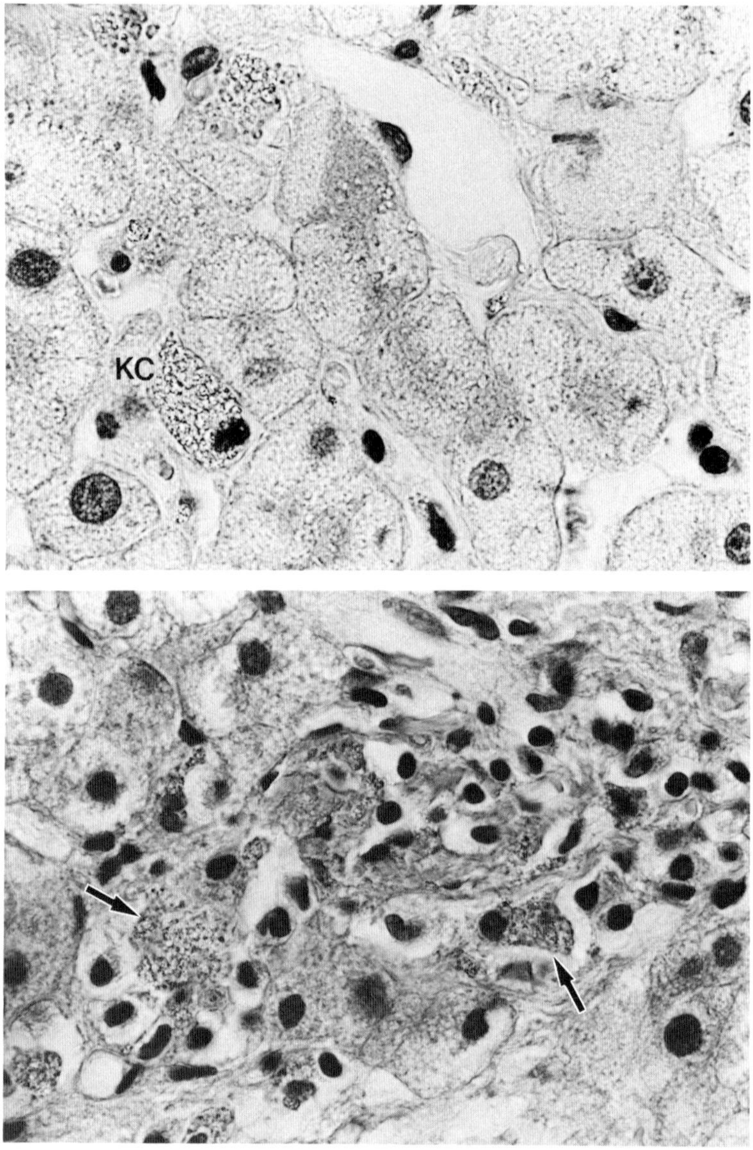

a

b

Fig. 3. Pigment deposits in the liver parenchyma during MTX treatment of RA revealed in liver biopsies. *a* Hepatocytes disclose some lipofuscin particles and show peribiliar basophilia. In a sinusoid, an enlarged Kupffer cell (KC) is loaden with pigment. HE × 290. *b* A small portal tract with pigment loaden macrophages (arrow) between lymphocytes. HE × 630.

Table IV. RA patients without MTX treatment

Patient	Sex	Age, years	RA activity	Total score	Special score
L.B.	F	62	2	13	3
M.B.	F	54	2	6	1
M.D.	F	35	3	6	1
H.E.	M	46	1	12	3
C.K.	M	70	3	6	2
K.K.	F	56	2	7	3
M.K.	M	43	1	9	5
M.P.	F	69	3	16	10
M.S.	F	52	3	8	4
H.S.	M	70	2	12	5

the alterations was considered, a high value corresponds to a more severe alteration. The total score summing up all variables is higher in the RA patients with MTX treatment than in the group without MTX therapy (12.2 and 9.5, respectively). The patients under MTX treatment disclosed a higher degree of those changes corresponding to an enhanced metabolism as well as higher content of pigment in the portal macrophages. There is no difference between both groups of patients when the score is restricted to those variables (special score) which are considered to be typical of MTX (variability of nuclear size, fatty change, necroses, portal tract inflammation, fibrosis, cirrhosis) according to the methotrexate guidelines [22]. The total and special scores are listed in table IV for RA patients without MTX application and in table V for RA patients with MTX therapy.

Discussion

Application of MTX even in low doses may be complicated by liver injury. Generally, two types of hepatotoxins can be differentiated [23]: Intrinsic toxins are those which regularly and predictably produce damage by direct or indirect mechanisms; liver damage of idiosyncratic hepatotoxins is not predictable, the suggested mechanism includes hypersensitivity.

MTX is among the intrinsic hepatotoxins. Fibrosis and even cirrhosis of the liver have been described as the most important altera-

Table V. RA patients with MTX treatment

Patient	Sex	Age years	RA activity	MTX application months	MTX dose, mg	Total score	Special score
H.H.	M	71	2	33	1,500	13	2
C.H.	F	74	1	33	1,500	16	4
D.D.	F	47	1	32	980	13	4
J.M.	F	58	2	35	2,890	10	2
E.M.	F	53	2	23	945	16	4
H.H.	F	65	2	26	1,500	15	6
A.G.	F	55	2	22	930	14	4
W.S.	M	64	1	24	450	10	3
H.E.	F	52	2	26	1,630	7	3
H.M.	F	46	3	31	1,320	10	2
G.G.	F	55	3	29	1,520	7	3
J.N.	F	64	1	21	1,060	17	4
F.P.	F	52	1	27	1,490	11	4
E.S.	F	50	3	45	1,920	13	5
J.H.	F	62	1	43	1,900	14	6
S.M.	F	49	2	4	280	13	5
K.A.	F	47	2	21	855	8	2

tions following MTX application. Their exact pathogenesis is still unknown, immunological phenomena do not seem to be of importance. HLA typing of psoriatic patients revealed different results [16]. *Zachariae* et al. [31] found an A_1, B_8 constellation of HLA antigens predisposing to the development of fibrosis or cirrhosis in MTX-treated psoriatic patients. From their light and electron microscopic studies on liver biopsies of MTX-treated psoriatics, *Nyfors and Poulsen* [15] concluded that piecemeal necrosis with a disruption of the limiting plate will be the first step towards liver damage. It was suggested that in a case of reduced regeneration capacity porto-portal and porto-central fibrous septa will develop and destroy the liver parenchyma. A similar pattern of liver injury had been described after treatment of acute leukemia in children using high dosage MTX [3]. Fibrosis of the liver did not develop in adults who had been treated with aminopterin or amethopterin because of solid tumors [24]. Regarding this finding, it has been concluded that the folic acid antagonist per se is not an inducer of liver fibrosis. Studies on MTX-treated psoriatics have shown that additional factors will promote the liver injury caused by MTX [5, 12, 13, 21]. Among those are the way of application, the cumulated MTX dosage, the patients' age, obesity and diabetes mellitus. Alcohol consumption is also an important additional factor [20, 27]. Finally, it has to be kept in mind that changes of the liver histology are found in psoriatics without MTX treatment more often than in age-matched control groups [20, 27].

In our study there was no patient with a completely normal liver histology. In both groups of patients the histological changes are qualitatively the same, but they differ in their degree. Hepatocellular and mesenchymal alterations can be differentiated. The epithelial changes consist of cytoplasmic swelling following fluid influx or proliferation of endoplasmic reticulum together with an increase in nuclear DNA content demonstrating an enhanced metabolic activity of the liver cells. Fatty degeneration of hepatocytes and single acidophilic liver cell necroses, which could be seen in a few patients' biopsies of both groups may be caused by toxic damage as well as other mechanisms. A basophilia of the hepatocellular cytoplasm in the region of the bile pole suggests a disturbed bile excretion [1]. The mesenchymal reaction is characterized by a proliferation of Kupffer cells and an infiltration of lymphocytes and monocytes in the portal tracts. Fibrosis of portal tracts which belongs to the mesenchymal reaction could also be observed only in a small number of patients. A third type of hepatic alteration consists of pigment deposition. Lipofuscin which is a residual product of intracellular lysosomal degradation may be increased with long-term intake of phenacetin [23]. Another type of pigment was observed in the Kupffer cells and portal tract macrophages. Its source is suggested in therapeutic compounds and their inert carrier substance. Regarding the higher amount of those pigments in the portal tract macrophages of MTX-treated RA patients, it is considered that MTX contributes to its formation.

The alterations which could be observed in the liver biopsies of both groups of patients are not specific. The histological changes of liver tissue are seen in the RA patients with and without MTX treatment with nearly the same frequency. Considering the degree of severity of these changes a small difference between RA patients without MTX and MTX-treated RA patients can be seen. For the patients without MTX therapy a score including all morphological variables of

9.8 and for the MTX-treated patients a score of 13.1 is calculated. But there is no different score if only those morphological variables are considered on which the classification of MTX damage according to the methotrexate guidelines [22] is based (fatty change, nuclear variability, portal inflammation, hepatocellular necroses, portal tract fibrosis, cirrhosis). Because of the missing significant correlation between cumulative MTX dosis and morphological score, MTX therapy is unlikely the determining cause of the described hepatic changes. Our study has confirmed the results of *Rau* [18], that in patients with rheumatoid arthritis treated with compounds other than MTX, changes of liver histology nearly always can be observed. They are a combination of reactive alterations in the course of rheumatoid arthritis and hepatotoxic effects of the therapeutic compounds. Morphologically, there is an overlap with residual alterations from antecedent liver diseases and changes in the liver related to other extrahepatic diseases or toxic injury. The demonstration of histological alterations of the same quality and severity in both groups of patients shows MTX not to be the only cause of these changes.

Conclusions

Liver biopsies of patients with treated rheumatoid arthritis (without MTX application) show reversible mesenchymal and hepatocellular changes of the liver histology. Alterations of the same quality and severity can be found in patients with MTX-treated rheumatoid arthritis.

These alterations can be caused by rheumatoid arthritis or its therapy (without MTX application), by MTX, by previous hepatic and extrahepatic diseases or various toxins. Because of the structural reaction pattern of the liver being limited, the importance of each factor cannot be defined.

In patients with MTX-treated rheumatoid arthritis the cumulative MTX dose does not correlate with the degree of histological alteration in the liver. We have not found evidence of severe liver injury related to MTX in the RA patients treated according to the reported schedule. However, liver biopsies prior to and during MTX application should be performed until these preliminary results have been confirmed based on a larger number of patients.

References

1 Altmann, H.W.: Drug-induced liver reaction: a morphological approach; in: Grundmann, Current topics in pathology, No. 69: Drug-induced pathology, pp. 69–142 (Springer, Berlin 1980).

2 Bergstresser, P.R.; Schreiber, S.H.; Weinstein, G.D.: Systemic chemotherapy for psoriasis. Archs Derm. *112:* 977–981 (1976).

3 Colsky, J.; Greenspan, E.M.; Warren, T.N.: Hepatic fibrosis in children with acute leukemia after therapy with folic acid antagonists. Archs Path. *59:* 198–206 (1955).

4 Dahl, M.G.C.; Gregory, M.M.; Scheuer, P.J.: Liver damage due to methotrexate in patients with psoriasis. Br. med. J. *i:* 625–630 (1971).

5 Dahl, M.G.C.; Gregory, M.M.; Scheuer, P.J.: Methotrexate hepatotoxicity in psoriasis – comparison of different dose regimens. Br. med. J. *i:* 654–656 (1972).

6 DiBartolomeo, A.G.; Mayes, M.D.; Bathon, J.M.: Methotrexate (MTX) in rheumatoid arthritis (RA). Arthritis Rheum. *27:* suppl., p. 61 (1984).

7 Edmundson, W.F.; Guy, W.B.: Treatment of psoriasis with folic acid antagonists. Archs Derm. *78:* 200–203 (1958).

8 Mackenzie, A.H.: Liver biopsy findings after methotrexate therapy for rheumatoid arthritis (Abstract). Clin. Pharmacol. Ther. *17:* 239 (1975).

9 Mackenzie, A.H.; Scherbel, A.L.: Management of rheumatoid arthritis in the surgical patient. Orthop. Clins. N. Am. *2:* 277–299 (1971).

10 Michaels, R.M.; Nashel, D.J.; Leonard, A.; et al.: Weekly intravenous methotrexate in the treatment of rheumatoid arthritis. Arthritis Rheum. *25:* 339–341 (1982).

11 Millward-Sadler, G.H.; Ryan, T.J.: Methotrexate induced liver disease in psoriasis. Br.J. Derm. *90:* 661–667 (1974).

12 Nyfors, A.: Liver biopsies from psoriatics related to methotrexate therapy. 3. Findings in post-methotrexate liver biopsies from 160 psoriatics. Acta pathol. microbiol. Scand., A, Pathol. *85:* 511–518 (1977).

13 Nyfors, A.: Methotrexate therapy of psoriasis. Dan. med. Bull. *27:* 74–96 (1980).

14 Nyfors, A.; Poulsen, H.: Liver biopsies from psoriatics related to methotrexate therapy. 2. Findings before and after methotrexate therapy in 88 patients. A blind study. Acta pathol. microbiol. scand., A, Pathol. *84:* 262–270 (1976).

15 Nyfors, A.; Poulsen, H.: Morphogenesis of fibrosis and cirrhosis in methotrexate-treated patients with psoriasis. Am. J. surg. Pathol. *1:* 235–243 (1977).

16 Nyfors, A.; Svejgaard, A.: The relation of HLA antigens to liver histology in methotrexate-treated psoriatics. Acta derm. vener., Stockh. *56:* 235–238 (1976).

17 Podurgiel, B.J.; McGill, D.B.; Ludwig, J.; Taylor, W.F.; Muller, S.A.: Liver injury associated with methotrexate therapy for psoriasis. Mayo Clin. Proc. *48:* 784–792 (1973).

18 Rau, R.: Die Leber bei entzündlichen rheumatischen Erkrankungen; in Schoen, Der Rheumatismus, vol. 43 (Steinkopff, Darmstadt 1978).

19 Robinson, J.K.; Baughman, R.D.; Auerbach, R.; Cimis, R.J.: Methotrexate hepatotoxicity in psoriasis. Archs Derm. *116:* 413–415 (1980).

20 Roenigk, H.H.; Bergfeld, W.F.; St. Jaques, R.; Owens, F.J.; Hawk, W.A.: Hepatotoxicity of methotrexate in the treatment of psoriasis. Archs Derm. *103:* 250–261 (1971).

21 Roenigk, H.H., Jr.; Auerbach, R.; Bergfeld, W.F.; Frost, P.; Maibach, H.; Tindall, J.; Weinstein, G.: A cooperative, prospective study of the effects of chemotherapy of psoriasis on liver biopsies; in Farber, Cox, Psoriasis. Proc. 2nd / Int. Symp., Stanford University, Stanford 1976, chap. 33, pp. 243–248.

22 Roenigk, H.H.; Auerbach, R.; Maibach, H.L.; Weinstein, G.D.: Methotrexate guidelines – revised. Am. Acad. Derm. *6:* 145–155 (1982).

23 Scheuer, P.J.: Liver biopsy interpretation; 3rd ed. (Baillière Tindall, London 1980).

24 Schoenbach, E.B.; Colsky, J.; Greenspan, E.M.: Observations on the effects of the folic acid antagonists, aminopterin and amethopterin, in patients with advanced neoplasms. Cancer *5:* 1201–1220 (1952).

25 Steinsson, K.; Weinstein, A.; Korn, J.; Abeles, M.: Low dose methotrexate in rheumatoid arthritis. J. Rheumatol. *9:* 860–866 (1982).

26 Weinstein, G.D.: Methotrexate. Ann. intern. Med. *86:* 199–204 (1977).

27 Weinstein, G.D.; Cox, J.W.; Supinga, W.T.; Millard, M.M.; Kalser, M.; Frost, P.: Evaluation of possible chronic hepatotoxicity from methotrexate for psoriasis. Archs Derm. *102:* 613 (1970).

28 Weinstein, G.; Frost, P.: Methotrexate for psoriasis. A new therapeutic schedule. Archs Derm. *103:* 33–38 (1971).

29 Wilke, W.S.; Calabrese, L.H.; Scherbel, A.L.: Methotrexate in the treatment of rheumatoid arthritis. Cleve. Clin. Q. *47:* 305–309 (1980).

30 Willkens, R.F.; Watson, M.A.; Paxson, C.S.: Low dose pulse methotrexate therapy in rheumatoid arthritis. J. Rheumatol. *7:* 501–505 (1980).

31 Zachariae, H.; Kragballe, K.; Thestrup-Pedersen, K.; Kissmeyer-Nielsen, F.: HLA antigens in methotrexate-induced liver cirrhosis. Acta derm. vener., Stockh. *60:* 165–166 (1979).

32 Zachariae, H.; Kragballe, K.; Søgaard, H.: Methotrexate induced liver cirrhosis. Studies including serial liver biopsies during continued treatment. Br. J. Derm. *102:* 407–412 (1980).

Prof. Dr. H. Frenzel, Pathologisches Institut der Universität Düsseldorf, Moorenstrasse 5, D-4000 Düsseldorf (BRD)

Rheumatology, vol. 9, pp. 235–237 (Karger, Basel 1986)

Discussion 9

Nyfors: The papers are now open for discussion.

Willkens: Dr. *Frenzel,* may I ask, do you have a classification for the 10 patients without MTX therapy the same way as you have for the 17 with MTX treatment?

Frenzel: Yes.

Willkens: Was there any patient in class 3 or 4?

Frenzel: No patient of both groups was in class 3 or 4, 65% were in class 1.

Furst: Dr. *Nyfors,* I was really quite interested in your putting together of the way in which cirrhosis occurs in MTX-treated patients, based on some inflammatory infiltrate. I had been told by our hepatologist that there really was no inflammatory component to the methotrexate effect on the liver. Dr. *Frenzl,* you are a pathologist? Is there inflammation as part of this or is there not?

Frenzel: Yes, we have seen in most patients, treated and untreated, mild infiltrations in the portal tracts. But we have never seen piecemeal necrosis which of course could be a result of the disease itself or another drug rather than methotrexate.

Furst: Is there evidence that methotrexate causes the sort of inflammation?

Frenzel: The data of our patients do not support the conclusion that methotrexate causes such an inflammation of the portal tracts. This inflammation was as frequently seen in the control group and with the same intensity.

Furst: Dr. *Nyfors,* you've had lots of experience, have you seen this specifically with methotrexate alone?

Nyfors: Well, it is seen in other diseases like chronic aggressive hepatitis, so the methotrexate sequence of histological events is a sort of a, I wouldn't call it an aggressive hepatitis, but it is more in terms of a persistent hepatitis, that we have seen in severe cases.

Rau: But piecemeal necroses are characteristic for chronic aggressive hepatitis and not for persistent hepatitis.

Furst: I was literally told by Dr. *Tolman:* forget the OT and PT, they are not going to help you.

Williams: Well, I think what Dr. *Tolman* was trying to say is that the OT and PT weren't predictive as to cirrhosis that the cirrhosis can occur without the presence of OT and PT and that you can get OT and PT elevations without the progression of cirrhosis.

Michaels: Perhaps we can ask the question in another way: Have you seen serial liver biopsies of patients with piecemeal necrosis that have not progressed to fibrosis or cirrhosis and have you seen serial biopsies, where you have not observed piecemeal necrosis and yet have observed cirrhosis?

Nyfors: Yes, we have seen piecemeal necrosis without progression to cirrhosis. On the other hand we have seen cirrhosis without previous piecemeal necrosis. But in most cases piecemeal necrosis is a prerequisite of cirrhosis. All together we have seen about 15–20 patients with cirrhosis developing during methotrexate therapy.

Michaels: Do you feel that piecemeal necrosis is an indication to stop methotrexate or have you continued it when you've seen piecemeal necrosis?

Nyfors: It depends a little on the number and how widespread they are. When you see the piecemeal necrosis in a lot of areas, we would make a pause with methotrexate therapy, and let the patient be without methotrexate for a year or so, and then perform another liver biopsy to see what happened.

Schmidt: The percentage of patients with cirrhosis after more than 5 years of MTX treatment which you have observed is rather high in comparison to rheumatoid patients. Would you think that the reason for it may be a specific predisposition of psoriatic patients?

Nyfors: No, I don't think they are specifically going to have cirrhosis. I think we were too enthusiastic about getting patients into methotrexate therapy in the past. So we got too many patients, and we didn't work hard enough on alcohol checks.

Dunky: Dr. *Nyfors,* did you see a difference in the frequency and severity of fibrosis and cirrhosis between i.v. and oral administration of MTX?

Nyfors: I always talk in my material about oral administration, I never use i.v., because the patients won't have that.

Willkens: I wonder if I could comment on Dr. *Schmidt's* question. I think there is some consideration that duration of time, not only dose, is predictive of an increased prevalence of fibrosis and cirrhosis. The English dermatologists have discussed that point and have done some control studies to establish that.

Weinstein: I'd like to pick up on what Dr. *Furst* has first been saying. Many hepatologists believe that the hepatocyte necrosis and inflammation seen with methotrexate is a separate and distinct process from the fibrosis seen with methotrexate. The fibrosis cannot be predicted by pathological changes of inflammation or hepatocyte necrosis, nor can it be predicted at all by transaminase elevations, which, in fact, correlate better with inflammation and piecemeal necrosis.

Nyfors: Well, I think, I agree. When I see transaminase elevations, I think of necrosis immediately, and not of fibrosis.

Dunky: A question to Dr. *Nyfors.* During the last years many prosiatic patients switched from MTX to PUVA. Have you any experience with patients who got in the past MTX and who are now on PUVA treatment, and have you made liver biopsies in these patients?

Nyfors: Well, of course, in my units a lot of methotrexate patients have been put on PUVA, and it seems that some of them have had elevations in the SGOT and so on. But they have not been liver biopsied. We've seen transaminase elevations, that's the only thing we have tested so far.

Karger: A question to Dr. *Nyfors:* Did you examine your patients with fibrosis and cirrhosis for hepatitis B antigen and antibody? Is there a higher incidence of hepatitis B in the history of patients with fibrosis and cirrhosis?

Nyfors: We did this routinely and I don't think we found any. And otherwise we would have excluded them from the group.

Rau: You saw piecemeal necrosis formation and bridging and you continued to treat these patients. Why didn't you stop the treatment before a complete cirrhosis could develop?

Nyfors: Well, in those days it was new, this piecemeal necrosis, and we didn't realize the importance of it. And we asked where we could, among hepatologists and pathologists, and we were advised just to go on, but do more frequent liver biopsies.

Weinstein: Dr. *Nyfors,* did any of those patients die from complications of their cirrhosis or fibrosis?

Nyfors: No, we are so lucky, and I talked on the telephone with *Zaccharia,* who has now about, I think, 700 liver biopsies, and that is from about 350 patients, and I have my 250 patients, and none of these patients have died from liver failure.

Karger: Dr. *Nyfors,* were your criteria for evaluation of liver biopsies in the 1970s the same as the revised criteria from 1982?

Nyfors: I think that our criteria were better, because we probably had one of the best pathologists in the world, Prof. *H. Paulsen,* to go through them. And we have done our own studies. We are not always compatible to the liver study group. But I don't think there is much disagreement between these studies, as the pathologists are usually working together.

Weinstein: Dr. *Nyfors,* just one thing, for the sake of argument, if you were to say that methotrexate is as toxic as you say in rheumatoid patients as in psoriatics, how many biopsies over what period of time with how much methotrexate will it take for us to find it, because we are not finding it.

Nyfors: Well, it will be interesting when you pass 4,000 mg.

Wilke: We are getting funded to do that kind of study right now. And our planning is to call back 100 patients who have had at least 10 years of therapy with methotrexate for rheumatoid arthritis, and who are still taking the drug and who apparently don't have any toxicity that we are aware of. And we will be biopsying those livers. Now, if we can get 100 of these patients, who have had 5 or 6 g of the stuff, maybe we can answer the question that, at least to say they are a subset of people with rheumatoid arthritis, who either can or cannot take the drug for that period of time.

Nyfors: I would say that the important thing is that you have either a multicenter study or a big unit having a lot of patients and then do a prospective study. So, I would advise pre- and postmethotrexate studies, and it can easily be done as multicentric.

Rheumatology, vol. 9, pp. 238–243 (Karger, Basel 1986)

Lung Complications of Methotrexate Therapy

E. K. Pöhler, M. Diekmann

Department of Medicine III, University of Cologne (Prof. *H.H. Hilger*), Cologne, FRG

Some cytostatic and immunosuppressive agents are known to be pneumotoxic, for instance azathioprine, BCNU, bleomycin, busulfan, chlorambucil, cyclophosphamide, cytosine arabinoside, melphalan, 6-mercaptopurine, methotrexate, mitomycin C, procarbazine and nitrogen mustard [1, 2].

Clinically drug-induced side effects of the lung can cause a slight fever, a dry cough, breathlessness at work or at rest and occasionally chest pain. In some cases, in addition to the pulmonary symptoms, erythema exsudativum multiforme can be observed. Laboratory tests show eosinophilia in about 40% of cases. Methotrexate-induced pneumonitis is a rare condition. In this condition an antibody could be demonstrated [3]. For this reason it seems to be likely that the side effect of the drug is allergic and induced by immunocomplex deposition. It is known that a precipitation of immunocomplexes can take place in pulmonary vessels. The destruction of these immunocomplexes by activation of the complement cascade results in damage to the lung tissue. This reaction is well known as type III allergy using the description of *Coombs and Gell* [4] and is one possible, but not proven, mechanism for methotrexate-induced pneumonitis. Airways obstruction following methotrexate therapy has not yet been described. The side effect may occur within 2 weeks after initiation of therapy (acute reaction), after months (subacute reaction) and after discontinuing therapy (chronic reaction). In most published cases the prognosis was good after discontinuing the drug and/or giving steroids.

It is important to know whether a drug reaction is of a facultative or an obligatory type in inducing pulmonary damage. This question can easily be answered with lung function equipment.

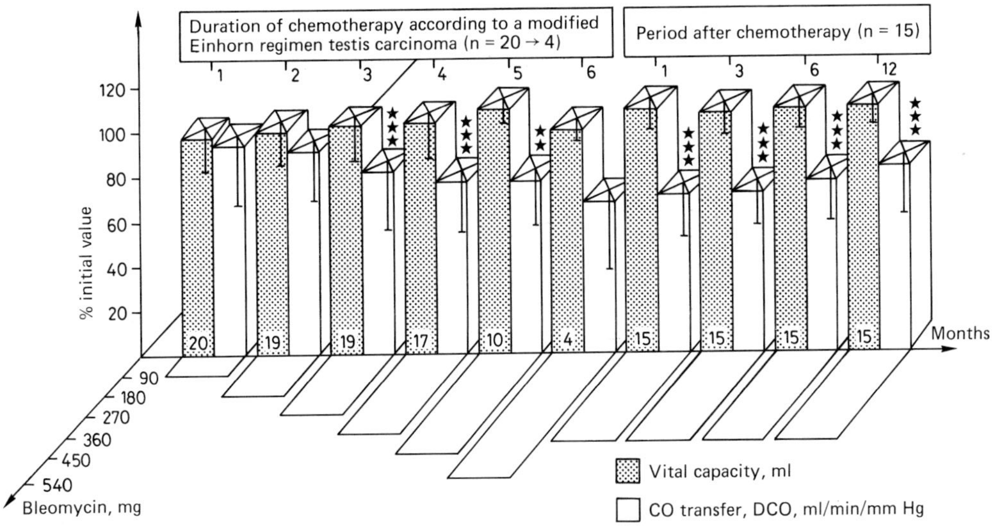

Fig. 1. Lung function before, during and after treatment with bleomycin (decrease in %).

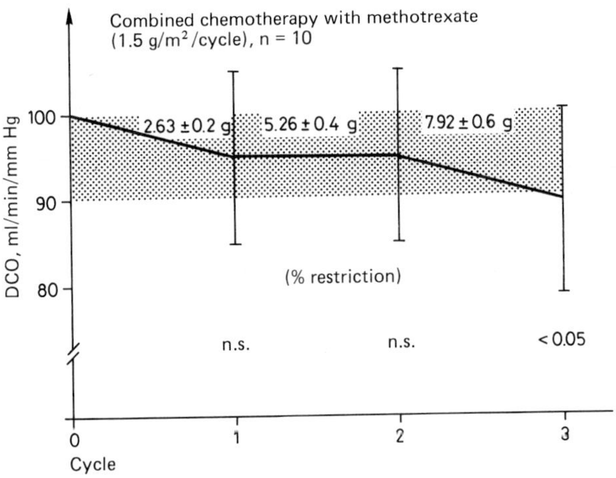

Fig. 2. Lung function before and during treatment with methotrexate in high dosage (decrease in %).

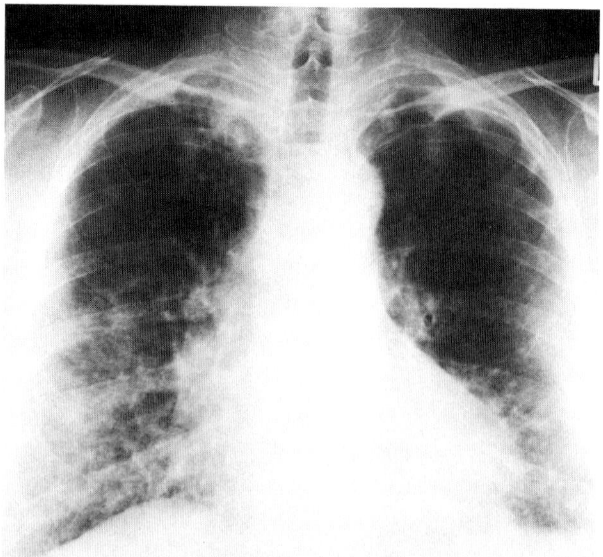

Fig. 3. Characteristic drug-induced alveolitis.

We prefer the carbon monoxide single breath method (DCO), when restrictive lung disease is to be investigated. The CO uptake is more sensitive than such volumetric methods as the vital capacity maneuver.

In all patients with a germ cell carcinoma treated with bleomycin a depression of lung function was shown using the DCO method (fig. 1). It takes about 1 year to restitute cytostatic-induced pulmonary impairment. That is why bleomycin is an obligatory drug inducing pulmonary damage. As we know the pharmacological basis of bleomycin toxicity is the generation of oxygen radicals. These molecules are cytotoxic and they show a preference for the lung tissue.

In a serial examination of lung function in patients with carcinoma of the breast treated with high dose methotrexate only a slight depression of DCO was visible. The total depression was about 10% (fig. 2). Statistically the result could not be proven for the result was within the range of error of the method. Although high dose therapy may cause pulmonary toxicity, no obligatory pulmonary damage as a

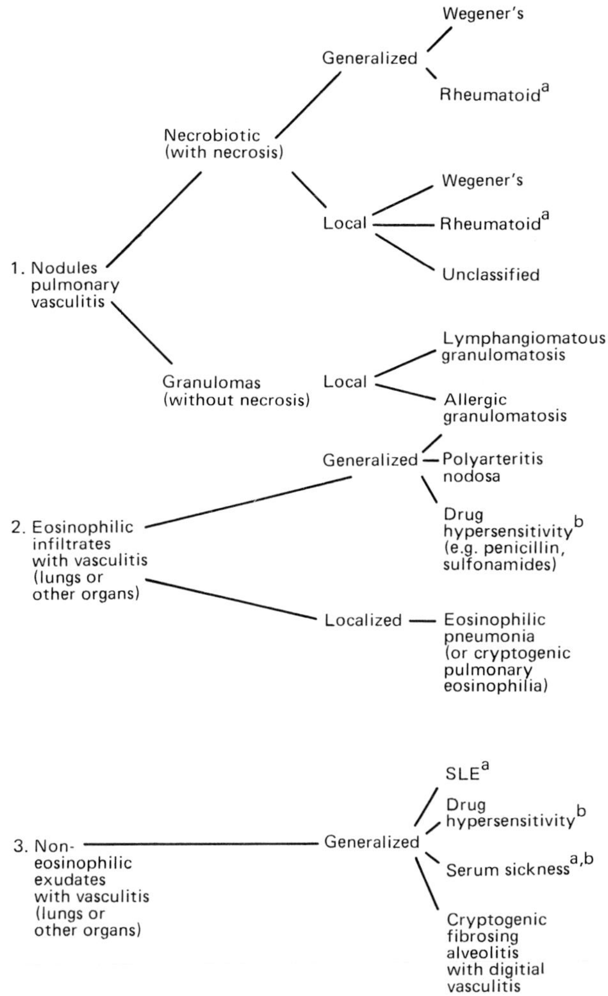

Fig. 4. Morphological changes of lung tissue in drug-induced disease. a = Easily demonstrable antibody; b = known antigen.

side effect of methotrexate could be demonstrated by any of the lung function tests. However, occasional unexpected lung complications have been reported. As this is a very rare complication and as in most of these cases a combination therapy was used it is very difficult to assume that MTX was the causative agent.

How can we evaluate drug-induced pneumonitis? A biopsy procedure is to be preferred to establish the diagnosis of methotrexate-induced side effects. Clinical methods and chest x-rays are unspecific. Further confusion is caused by many opportunistic infections which show patterns of interstitial pneumonia, typical for methotrexate-induced pneumonitis.

Figure 3 demonstrates a characteristic drug-induced alveolitis with bilateral infiltration of the lung showing a preference for the middle and lower parts of the lung. The confluence of reticular-nodular pattern can be seen in widespread areas. With such an X-ray it is possible to confirm the diagnosis, but in the advanced stages of pneumonitis the changes shown in the X-ray are no longer so specific and are open to differential diagnosis. In particular in methotrexate-treated patients who are immunosuppressed, nosocomial and opportunistic infections must be considered. In many cases the differentiation between infectious and toxic changes by X-ray investigation is impossible.

For the differential diagnosis of infectious pulmonary infiltrations enteric self-infection with *Staphylococcus aureus,* candida, cytomegalic virus, herpes virus and *Pneumocystis carinii* must be considered. To establish the diagnosis the method of choice is fiberoptic forceps biopsy.This procedure is not dangerous and tolerable even for dyspnoic patients. The histology gives a good description of the changes in lung tissue. In drug-induced damage, vasculitis is characteristic, combined with granulomas, eosinophilia or nonspecific cell infiltration [5]. The histological pattern can also differentiate between changes seen in the rheumatic diseases and typical drug-induced alterations. The combination of biopsy with bronchoalveolar lavage and a bacteriological culture allows a diagnosis in most cases (fig. 4).

Conclusion

Methotrexate pneumonitis is rare. Up till now no obligatory pneumotoxic side effect has been proven. If a methotrexate-induced pneumonitis is suspected a forceps biopsy using a bronchoscope combined with bronchoalveolar lavage and a bacteriological culture examination is indicated. This procedure will establish a diagnosis and decide the appropriate therapy.

References

1 Ginsberg, S.J.; Louis, R.L.: The pulmonary toxicity of antineoplastic agents. Semin. Oncol. *1:* 34–51 (1982).

2 Sostman, H.D.; Morthay, R.A.; Putman, C.E.: Cytotoxic drug-induced lung disease. Am. J. Med. *62:* 608–613 (1977).

3 Da Costa, M.; Isacoff, W.; Rothenberg, S.P.; Igbal, M.P.: Protein-methotrexate-IgG complexes in the serum of patients receiving high-dose antifolate therapy. Cancer *46:* 471–474 (1980).

4 Coombs, R.R.A.; Gell, P.G.H.: Classification of allergic reactions responsible for clinical hypersensitivity and disease; in Gell et al., Clinical aspects of immunology; 3rd ed., p. 761 (Blackwell, Oxford 1968).

5 Turner-Warwick, M.: Relations between clinical immunology and respiratory diseases: necrotizing angiitis with lung involvement. Therapiewoche *26:* 8358–8367 (1976).

Dr. Eike K. Pöhler, Department of Medicine III, University of Cologne,
Joseph-Stelzmann-Strasse 9, D-5000 Cologne 41 (FRG)

Rheumatology, vol. 9, pp. 244–251 (Karger, Basel 1986)

Methotrexate-Induced Pulmonary Toxicity in Patients with Rheumatoid Arthritis

H. James Williams, Grant W. Cannon, John R. Ward
University of Utah School of Medicine, Salt Lake City, Utah, USA

Toxicity to methotrexate (MTX) therapy is generally manifested by stomatitis, bone marrow suppression, skin rashes, and hepatic fibrosis. Less well recognized is acute pneumonitis associated with MTX which was reported in 1969 in patients with acute lymphoblastic leukemia [1, 2]. This acute lung disease was later described in patients treated with MTX for solid tumors [3], psoriasis [4] and other skin diseases [5], and polymyositis [6].

An additional form of pulmonary toxicity to MTX, pleurisy, has also been described in 14 patients treated with MTX for trophoblastic tumors [7]. Four of these patients also had pleural effusions or localized lung collapse but none developed parenchymal infiltrates.

MTX has recently been recommended for treatment of refractory rheumatoid arthritis (RA) [8, 9]. As experimental use of MTX in RA has increased, reports of acute pneumonitis resulting from MTX have appeared [10, 11].

There is no reported correlation of MTX-induced pneumonitis with age, sex, or disease [12]. Lung disease has occurred whether the MTX was given orally [1, 4], intramuscularly [1], intravenously [6], or intrathecally [3]. It does not appear to be related to daily, weekly, or total dose. Total doses as small as 145 mg [11] have resulted in lung disease and pneumonitis has first been noted after 35,000 mg [13]. While it was initially thought that lung toxicity only occurred on doses of at least 20 mg/week [14], pneumonitis has been described in RA patients receiving only 7.5 mg/week [10].

Table I. Methotrexate-induced lung disease in patients with RA

Methotrexate therapy						Reference
age	sex	weeks	weekly dose	cumulative dose	outcome	
75	F	54	15	765	recovered	15
61	F	51	7.5	383	recovered	15
63	M	30	7.5	225	dyspnea	15
62	M	24	15	360	recovered	10,15
66	F	20	7.5	165	recovered	10,15
55	M	16	7.5–10	145	recovered	11
–	–	8	7.5–12.5	–	recovered	16
–	–	8		–	recovered	16
–	–	52		–	died	16

Prevalence

The original report of MTX-induced lung disease identified 38 episodes of pneumonitis in 93 children with acute lymphoblastic leukemia taking MTX [2]. However, this study certainly overestimated the true prevalence of MTX-induced pulmonary toxicity and contained at least 1 patient with *Pneumoncystis carinii* pneumonia. The incidence of pneumonitis in patients taking MTX is unknown.

The frequency of pulmonary toxicity in RA patients receiving MTX is difficult to determine but has been estimated at 4% in one survey of 127 patients with RA [15]. However, in another study of 189 patients taking low-dose pulse MTX or placebo for 18 weeks for RA, no pneumonitis was noted in the 95 patients taking MTX [8]. There are 9 reported cases of MTX-associated pulmonary toxicity in patients with RA. Three cases are well described [9, 10] and 6 cases are presented in abstract form [15, 16]. Table I outlines the age, sex, methotrexate therapy, and outcome in these patients. Total MTX dose has varied from 145 to 765 mg and was given as low-dose pulse therapy of 7.5–15 mg/week. Seven patients have completely recovered and 1 patient had some residual dyspnea after 1 month's follow-up. One patient died of respiratory failure attributed to MTX.

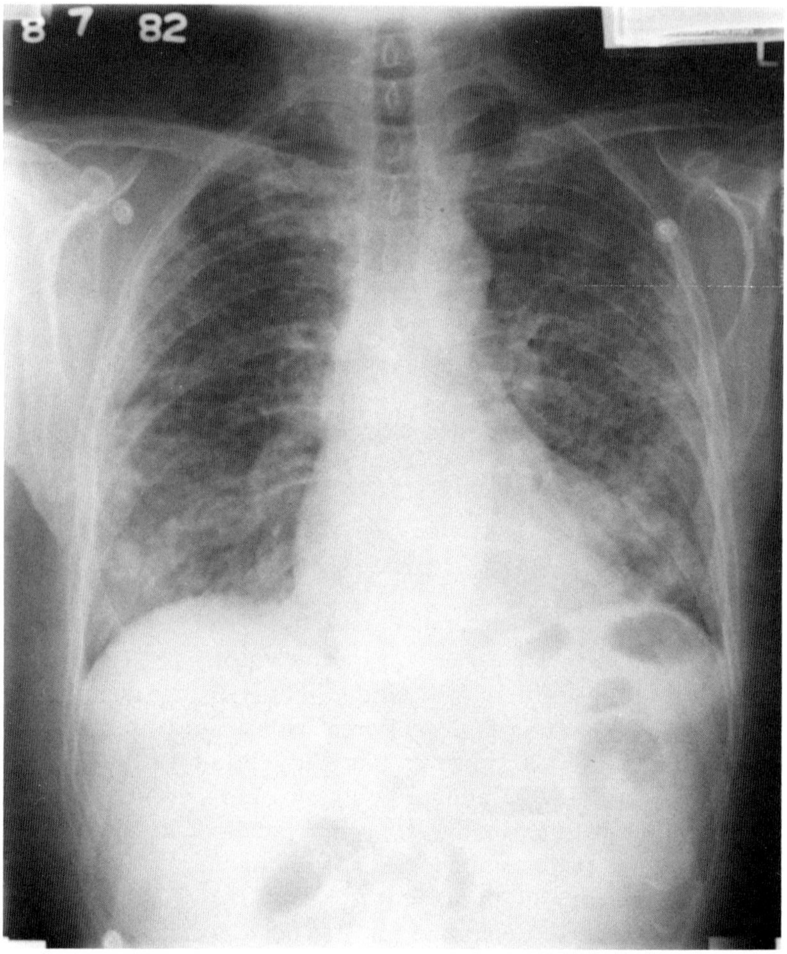

Fig. 1. Chest radiograph of a patient with methotrexate-induced pneumonitis indicating bilateral interstitial infiltrates in bases and mid-lung fields. Reproduced with the permission of the authors [10].

Clinical Presentation

The clinical features of MTX pulmonary toxicity in RA are similar to those which have been well described in other diseases [12, 14, 17]. Headache and malaise are frequent prodromal symptoms followed by dyspnea and a dry cough. Fever is almost always present and patients are often tachypneic. Cyanosis was reported in approximately half of the patients [14] and has also been seen in 1 RA patient [10]. About one-third of the patients develop crepitant rales [14]. Peripheral blood eosinophilia is reported in about half of the patients [12].

Arterial blood gases usually demonstrate hypoxemia and hypocapnea [12, 14]. There is often a decreased diffusion capacity and a restrictive ventilatory defect may be seen. Persistent reduction in vital capacity and in diffusion capacity have been reported in patients with apparent clinical recovery [14]. *Wall* et al. [18] prospectively studied lung function with spirometry, plethysmography, and diffusion capacity to carbon monoxide in patients receiving high dose MTX. They concluded that no pulmonary changes were apparent in these adolescent patients with total MTX doses of up to 256 g/m^2.

Chest roentgenograms initially reveal interstitial changes and edematous interlobular septa. These changes are generally bilateral and occur predominantly in the bases or mid lung fields (fig. 1). The diffuse interstitial infiltrates may progress into alveolar disease with confluent densities and air bronchograms [19]. Segmental or lobar consolidations have not been reported [12]. Radiographic changes clear rapidly with cessation of MTX therapy although chronic fibrosis has been described [13, 14, 20].

Diagnosis

The diagnosis of MTX-induced pneumonitis in patients with RA is as difficult as it is in patients with malignancies, other connective tissue disease, and/or immunosuppression. Interstitial inflammation and fibrosis can result as a complication of RA. The pulmonary symptoms and radiographic findings may also be a result of leukemic infiltrate, metastasis, and/or viral or opportunistic infection, particularly with *carinii. Johnson and Johnson* [21] performed needle aspiration in pulmonary disease complicating chemotherapy in children and found *P. carinii* in 11 patients taking MTX. Even concomitant drug therapy may produce a similar clinical picture. Patients have been described with assumed MTX pneumonitis that have also been receiving bleomycin [20] or nitrofurantoin [10], either of which can result in interstitial pneumonitis. Because of these additional possible causes of interstitial

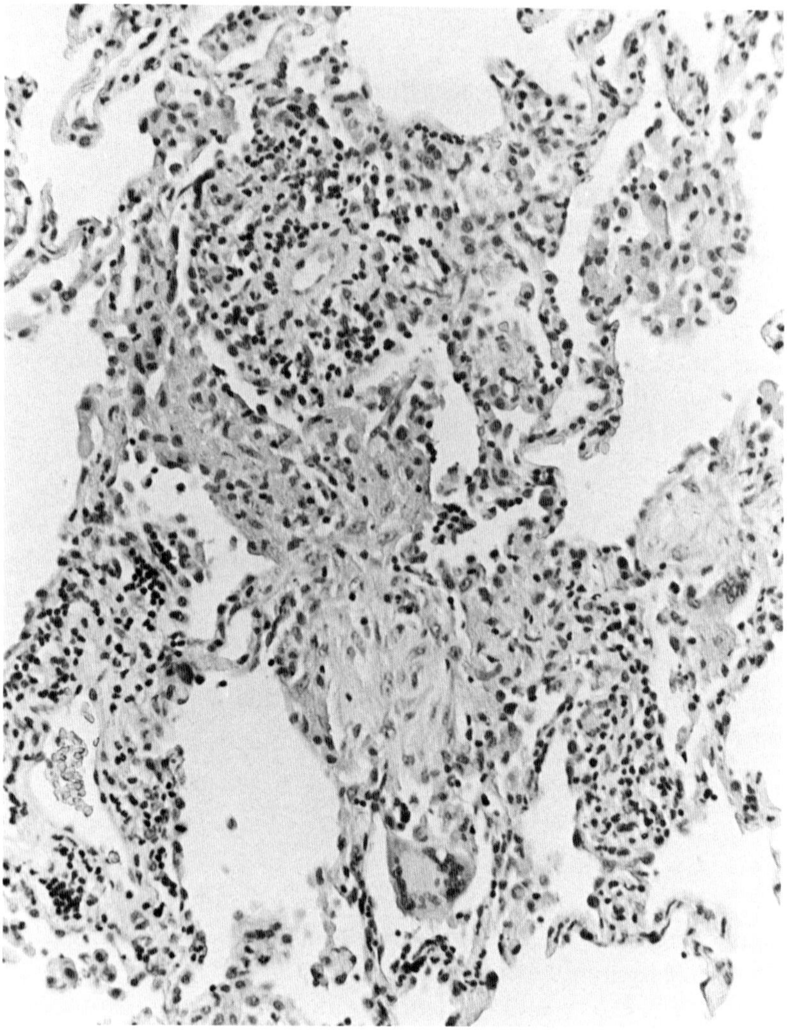

Fig. 2. Histological specimen from pulmonary biopsy of a patient with methotrexate-induced lung disease demonstrating inflammatory cells, multinucleated giant cells and bronchiolitis. Reproduced with the permission of the authors [10].

pulmonary disease, open lung biopsy is often necessary to fully evaluate the patient.

Histopathologic examination of pulmonary tissue (fig. 2) resembles a hypersensitivity pneumonitis [22] with an interstitial granulomatous reaction, giant cells, and even bronchiolitis [10, 14]. Aggregates of mixed inflammatory cells are present in the alveoli and in the interstitium. Lymphocytes predominate but plasma cells and eosinophils are also seen. Diffuse alveolar damage with hyaline membranes and atypical alveolar cells are described [14] and suggest a toxic drug reaction. Noncaseating granulomas with large giant cells have been reported [4, 10, 19] and bronchiolitis obliterans is occasionally noted [10]. Varying degrees of fibrosis may be present but it is usually mild [10, 21]. Progressive interstitial fibrosis is an uncommon result, but has been described [13, 20, 23].

Electron microscopy reveals desquamation and proliferation of macrophages, interstitial infiltrates with lymphocytes and eosinophils, and intact basement membranes [14].

We propose the following criteria for making a diagnosis of MTX-induced pulmonary toxicity: (1) There should be exposure to MTX. The total dose, frequency, or length of therapy is irrelevant. (2) The clinical course should resemble a hypersensitivity reaction. (3) There should be exclusion of other explanation of pulmonary disease, particularly excluding the underlying disease process and infectious agents and other drugs with similar pulmonary toxicity. (4) Lung pathology should be consistent with drug-induced toxicity.

The etiology of the MTX-induced lung disease is unknown. The peripheral eosinophilia and granulomatous pneumonitis with giant cells are consistent with a hypersensitivity reaction but patients have continued taking MTX with complete resolution of their pulmonary problems [1]. Direct toxicity to the lung by MTX has also been suggested but evidence to support this supposition is lacking. Infection could also be implicated; however, cultures done from bronchial washings and biopsy tissues have generally been negative [1, 10, 17].

Treatment

Previous reports document resolution of the pulmonary symptoms and the radiographic infiltrates following withdrawal of the MTX, usually within weeks. However, resolution has also been noted when patients continued to receive MTX [1]. Steroids have been advocated [12], although patients have improved without steroid therapy [1]. It has been suggested that the use of corticosteroid hastens resolution of the disease process [14]. Three patients had rapid clearing of pulmonary infiltrates when treated with daunorubicin [24].

Death from MTX pneumonitis is uncommon and has usually occurred in patients who have not received corticosteroid therapy [12]. One of the 9 patients with RA died from pulmonary insufficiency despite treatment with corticosteroids [16]. A few patients, including 1 with RA [15], have been rechallenged with MTX without recurrence of the lung disease, although one patient with mycosis fungoides was retreated with recurrence of pulmonary infiltrates [5]. The development of chronic interstitial fibrosis is rare but has been reported [13, 20].

Conclusions

MTX-induced pulmonary toxicity in RA patients is similar to that reported in other diseases. It can occur on any dose, by any route of administration, in any disease. The onset of symptoms is insidious with fever, dyspnea, and dry cough nearly always present. Withdrawal of MTX usually results in improvement of symptoms which may be hastened by treatment with corticosteroids. Patients have been successfully rechallenged with MTX following resolution of the lung disease without recurrence of symptoms. Death or chronic fibrosis can occur but are not common. However, MTX-induced pneumonitis is a potentially serious complication and evaluation of pulmonary complaints in patients taking MTX is essential.

References

1 Clarysse, A.M.; Cathey, W.J.; Cartwright, G.E.; Wintrobe, M.M.: Pulmonary disease complicating intermittent therapy with methotrexate. J. Am. med. Ass. *209:* 1861–1864 (1969).

2 Acute Leukemia Group B: Acute lymphoblastic leukemia in children. Maintenance therapy with methotrexate administered intermittently. J. Am. med. Ass. *207:* 923–928 (1969).

3 Gutin, P.H.; Green, M.R.; Bleyer, W.A.; Bauer, V.L.; Wiernik, P.H.; Walker, M.D.: Methotrexate pneumonitis induced by interthecal methotrexate therapy. A case report with pharmacokinetic data. Cancer *38:* 1529–1534 (1976).

4 Filip, D.J.; Logue, G.L.; Harle, T.S.; Farrer, W.H.: Pulmonary and hepatic complications of methotrexate therapy of psoriasis. J. Am. med. Ass. *216:* 881–882 (1971).

5 Goldman, G.C.; Moschella, S.L.: Severe pneumonitis occurring during methotrexate therapy. Report of two cases. Archs Derma. *103:* 194–197 (1971).

6 Arnett, F.C.; Whelton, J.C.; Zizic, T.M.; Stevens, M.B.: Methotrexate therapy in polymyositis. Ann. rheum. Dis. *32:* 536–546 (1973).

7 Walden, P.A.M.; Mitchell-Heggs, P.F.; Coppin, C.; Dent, J.; Bagshawe, K.D.: Pleurisy and methotrexate treatment. Br. med. J. *ii:* 867 (1977).

8 Willkens, R.F.; Watson, M.A.; Paxson, C.S.: Low dose pulse methotrexate therapy in rheumatoid arthritis. J. Rheumatol. *7:* 501–505 (1980).

9 Ward, J.R.: A prospective randomized controlled trial of low dose pulse methotrexate in rheumatoid arthritis. Arthritis Rheum. *27:* suppl. 4, p. S40 (1984).

10 Cannon, G.W.; Ward, J.R.; Clegg, D.O.; Samuelson, C.O., Jr.; Abbott, T.M.: Acute lung disease associated with low-dose pulse methotrexate therapy in patients with rheumatoid arthritis. Arthritis Rheum. *26:* 1269–1274 (1983).

11 Englebrecht, J.A.; Calhoon, S.L.; Scherrer, J.J.: Methotrexate pneumonitis after low-dose therapy for rheumatoid arthritis. Arthritis Rheum. *26:* 1275–1278 (1983).

12 Ginsberg, S.J.; Comis, R.L.: The pulmonary toxicity of antineoplastic agents. Semin. Oncol. *9:* 34–51 (1982).

13 Kaplan, R.L.; Waite, D.H.: Progressive interstitial lung disease from prolonged methotrexate therapy. Archs Derma. *114:* 1800–1802 (1978).

14 Sostman, H.D.; Matthay, R.A.; Putnam, C.E.; Smith, G.J.W.: Methotrexate-induced pneumonitis. Medicine *55:* 371–388 (1976).

15 Cannon, G.W.; Clegg, D.O.; Samuelson, C.O., Jr.; Ward, J.R.: Pulmonary toxicity during treatment of rheumatoid arthritis with methotrexate. Prevalence, clinical features, treatment, and follow up. Arthritis Rheum. *27:* suppl. 4, p. S26 (1984).

16 StClair, E.W.; Rice, J.R.; Snyderman, R.: Severe pneumonitis complicating low-dose oral methotrexate therapy in rheumatoid arthritis. Arthritis Rheum. *27:* suppl. 4, p. S60 (1984).

17 Rosenow, E.C., III: The spectrum of drug-induced pulmonary disease. Ann. intern. Med. *77:* 977–991 (1972).

18 Wall, M.A., Wohl, M.E.B.; Jaffe, N.; Strieder, D.J.: Lung function in adolescents receiving high dose methotrexate. Pediatrics, Springfield *63:* 741–746 (1979).

19 Everts, C.S.; Westcott, J.L.; Bragg, D.G.: Methotrexate therapy and pulmonary disease. Radiology *107:* 539–543 (1973).

20 Bedrossian, C.W.M.; Miller, W.C.; Luna, M.A.: Methotrexate-induced diffuse interstitial pulmonary fibrosis. South. med. J. *72:* 313–318 (1979).

21 Johnson, H.D.; Johnson, W.W.: *Pneumocystis carinii* pneumonia in children with cancer. Diagnosis and treatment. J. Am. med. Ass. *214:* 1067–1073 (1970).

22 Katzenstein, A.A.; Askin, F.B.: Surgical pathology of non-neoplastic lung disease, p. 33 (Saunders, Philadelphia 1982).

23 Nesbit, M.; Krivit, W.; Heyn, R.; Sharp, H.: Acute and chronic effects of methotrexate on hepatic, pulmonary, and skeletal systems. Cancer *37:* 1048–1054 (1976).

24 Pasquinucci, G.; Ferrara, P.; Castellari, R.: Daunorubicin treatment of methotrexate pneumonia. J. Am. med. Ass. *216:* 2017 (1971).

H. James Williams, MD, University of Utah Medical Center,
50 North Medical Drive, Salt Lake City, UT 84132 (USA)

Rheumatology, vol. 9, pp. 252–253 (Karger, Basel 1986)

Discussion 10

Jürgens: Both, Dr. *Pöhler's* and Dr. *Williams',* papers are open for discussion. Are there any questions?

Weinstein: Dr. *Williams,* someone in your group made a suggestion that lung toxicity may have an incidence as high as 4%. You still think that that is a valid percentage?

Williams: We saw 5 cases in a matter of months. So we then went back and looked at all the patients we had treated, that was 127 patients who had received methotrexate; from that data they concluded a proposed prevalence of 4%. Since that time we have continued to treat patients with methotrexate, but had no further cases of methotrexate pneumonitis.

Karger: Dr. *Williams,* can you state clearly, or do I have to ask the pathologist, that the histological changes you mentioned in the MTX-induced pneumonitis are distinct or different from the changes that you can find in the so-called rheumatic lung?

Williams: I would like to defer to Dr. *Frenzel* on pathological items.

Frenzel: What was shown in your slides looks like an extrinsic alveolitis according to a hypersensitivity reaction. I am not aware whether one can find similar alterations in rheumatoid arthritis, I know of lung alterations with fibrosis of the alveolar walls with infiltrations of lymphocytes and plasma cells, but not such a distinct granuloma formation. And I would think that the shown granuloma epithelioid cells and giant cells of the Langhans type are indicating an extrinsic hypersensitivity alveolitis.

Williams: That was the impression of our pathologists also, and all of our patients have appeared to have a hypersensitivity-like reaction.

Jürgens: Dr. *Williams,* as far as I am aware, there are very few reports in the literature of anaphylactic reactions occurring in patients receiving methotrexate. And there has been one report in 'Cancer', if I remember correctly it was in 1981, where immune complexes were detected in patients. Are you aware of any studies that immune-complex deposition has been detected in lung tissue?

Williams: Only when Dr. *Pöhler* has mentioned it today, I have not seen any evidence or any articles referring to immune complexes before.

Pöhler: It is a very difficult question, because elimination of immune complexes is very fast in the lung, and I think that you must have a biopsy within 12 up to 24 h, after that time you won't find any immune complex.

Jürgens: Any other questions?

Michaels: Dr. *Williams,* I think you mentioned that you have 1 patient with methotrexate whom you felt had a methotrexate lung disease whom you re-challenged with methotrexate. Was it a mild syndrome he had that you felt it was not dangerous to re-challenge or was there another possibility of infection or farmer's lung or whatever?

Williams: Those patients with methotrexate lungs have been sick. They have had shortness of breath, cough hypoxemia, but they recovered fairly rapidly, and this patient recovered completely. So since the previous litrature had indicated that some of these patients recover spontaneously, and others have been re-challenged in other diseases, we placed them back on methotrexate and up to this point, I think it has been 4 months now, they've had no problems. Only the 1 patient with mycosis fungoides has re-developed symptoms.

Willkens: Do you think transbronchial biopsies are adequate to assess interstitial change secondary to these drugs? Transbronchial versus open biopsies.

Pöhler: I think it is of the same value as open lung biopsy and it has not the same risk, and you can, if you do not get the result in the first approach, repeat it, but then I am sure you will have the result.

Weinstein: I have a question for both, actually. From the data, one could question whether or not methotrexate induced pulmonary toxicity. At least from Dr. *Williams'* data, only 1 patient got it again on re-challenge. And since, epidemiologically, all 5 cases occurred within a short period of time, and he has not seen another case since, does he really believe that methotrexate is the cause of this lung toxicity?

Williams: The 5 cases that occurred in our institution, for these there was no other explanation. I do not know that we understand methotrexate pneumonitis, but yet, I think it occurs.

Furst: I think on a strictly clinical pharmacological basis, if I re-challenge, and it does not occur, you cannot ascribe it to the drug.

Karger: Three questions to Dr. *Williams.* First: How long does it take to develop MTX-induced pneumonitis? Second: Did you find any bacteria in the bronchial lavage? Third: Did your patients receive antibiotics additionally to the corticosteroids?

Williams: All our patients had negative cultures. I think the first one was treated with antibiotics without response, but after that none of the others were treated with antibiotics. The lowest total dose to develop MTX-induced toxicity was 145 mg.

Dunky: Has HLA typing been performed in your patients?

Williams: These patients have not been HLA typed.

Wilke: I have a question for Dr. *Rau:* How many patients have you treated with methotrexate here?

Rau: About 140 patients now.

Wilke: Alright, and in what period of time have you been treating these patients?

Rau: The longest period is about 4 years.

Wilke: O.K. Have you experienced this lung toxicity?

Rau: No, I wanted to present 2 cases today, but there was not enough time. In both cases it was possible to find another explanation for these lung infiltrations than a side effect of MTX therapy.

Wilke: The reason I ask is that we have seen 1 case too and that was someone with 35 mg of methotrexate for polympositis. And it is really quite dramatic, and frightening. And this patient did recover rather nicely and quickly. But we have not seen it with a low dose.

Williams: I certainly cannot argue with *Furst* that when we cannot re-challenge them and find it, that it makes it somehow difficult to make this diagnosis. But it has been described in so many different conditions by so many different people that I wonder if there is not a real association, but I cannot explain it.

Furst: I believe that it must happen once in a while, because you do see it. But I suspect that the incidence is going to be lower than what we have now.

Rheumatology, vol. 9, pp. 254–258 (Karger, Basel 1986)

Round Table Discussion

Rau: In this round table I wish to ask some questions to different aspects of MTX treatment in rheumatic diseases, and as our time is short, I ask you to give very short statements. The first section will be the mode of action. Dr. *Jürgens,* would you please summarize the mode of action of MTX on tumor cells.

Jürgens: A short statement would be: S phase arrest.

Rau: Dr. *Nyfors:* What is the mode of action in psoriasis and in psoriatic arthritis?

Nyfors: To the first question I can be similarily short: S phase arrest and the second question I can be even shorter: I don't know.

Rau: Dr. *Dunky,* would you agree?

Dunky: Yes, we don't know really how it is acting in psoriatic arthritis. My impression is, that we can afford longer intervalls or lower doses in psoriatic arthritis than in rheumatoid arthritis, but I think that depends on the severity of the disease.

Rau: The next question was to Prof. *Miescher,* who is no longer with us. I wanted to ask him, if he can explain the good results observed in his patients with autoimmune diseases by an immunosuppressive effect of methotrexate. According to his observations the effect lasts about 10 days, there after he had to repeat the dose. Dr. *Willkens,* can you tell us, which mechanisms of action are discussed in rheumatoid arthritis and if these mechanisms are more symptomatic or more remittive?

Willkens: The answer is: I don't know. I would speculate that there may be a dual effect, both, if you call symptomatic anti-inflammatory as well as a more prolonged effect on the impairment of the immune apparatus as well.

Rau: So we have to state that our knowledge about low-dose MTX treatment in RA is still very limited. Now some questions concerning the effectivity of MTX: Is MTX the most effective drug in the treatment of psoriasis and is that also true for psoriatic arthritis, Dr. *Nyfors?*

Nyfors: I can't answer for psoriatic arthritis. In my patients the pain disappeared in one-third, in one-third the pain was diminished and one-third did not respond. That is all I can say about it. Regarding skin psoriasis, which is probably of the least interest here, we have alternative treatments like PUVA, psoralen and UVA light and we have these different retinoids. So, methotrexate is one of three very good treatments for skin psoriasis.

Rau: So, you would say, these treatments are as effective as methotrexate in psoriasis?

Nyfors: Yes.

Rau: And Dr. *Dunky,* what do you think, is MTX the most effective treatment of psoriatic arthritis or not?

Dunky: We use in psoriatic arthritis gold regularly and MTX, and to our experience methotrexate is the most effective drug.

Rau: We have heard that also in Reiter's syndrome, in the case of arthritis as well as regarding skin and nail changes, methotrexate is a very effective form of treatment. Now, I wish to ask Prof. *Gross,* you made direct comparisons in rheumatoid arthritis with patients who were treated with methotrexate for a certain period of time, and then switched to azathioprine or 6-mercaptopurine. And what was your impression when comparing these treatments?

Gross: Well, we had a good response to MTX and to azathioprine. I have some relapses with azathioprine and 6-mercaptopurine. Mercaptopurine is weaker than azathioprine. And I feel that MTX was a bit weaker than azathioprine, too. But we stopped completely the methotrexate medication, maybe after too short a time. As you heard this morning, we have administered MTX only for 20 weeks. We stopped MTX therapy completely thereafter.

Rau: As far as I know, there are no or only a few short time studies to compare the efficacy of methotrexate with other cytotoxic agents. Dr. *Wilke,* what comparative studies should be undertaken to further evaluate the position of MTX in the treatment of RA?

Wilke: I think, if we are going to compare cytotoxic agents, the only other agent that I would consider at this stage of the game and with what we know about alkylating agents, is azathioprine. And I suspect that a controlled study, comparing methotrexate to azathioprine, might be of some value. When we look at the controlled studies that have been done already with azathioprine, I think we might predict that methotrexate will have a somehow faster onset of action, but I don't know, I can't be sure that patients 6 months down the line would be better off on one or the other drug. I think that would be very interesting to see that sort of comparison. Now, let's say something about where we are right now. Some of us, me, are beginning to wonder whether methotrexate might not be the drug to choose even at the same time that some other people would be thinking about *D*-penicillamine, maybe even gold, so that somewhere down the line comparisons both from the standpoint of toxicity and efficacy need probably to be done comparing gold, *D*-penicillamine and methotrexate. It may be, 10 years down the line, people will be using methotrexate about the same time that people now use gold and *d*-penicillamine. Maybe, I don't know that, but I would like to see those trials, too.

Rau: Let us move on to the indications of MTX. As you said, the patients who are put on MTX need not be, in every case, treated with gold and *D*-penicillamine before starting with MTX. So I ask you once more, when do you think MTX is indicated in rheumatoid arthritis patients, Dr. *Wilke*?

Wilke: My personal opinion is that methotrexate is indicated when patients have failed other reasonable conservative therapy. Other reasonable conservative therapy should certainly include high-dose, nonsteroidal anti-inflammatory drugs. It should certainly include some sort of remittive agents, gold, penicillamine or chlorochine. Probably, it should include low-dose corticosteroids, although you could argue about that. At that point, I think it becomes legitimate to begin to wonder: Do I use gold, *D*-penicillamine, methotrexate? You have to individualize that to the patient.

Rau: So you would say, that MTX is a true alternative to gold and *D*-penicillamine?

Wilke: It is at our institution. But that's a minority opinion.

Furst: I would like to put in a word of caution here. As Dr. *Wilke* mentioned, it is possible that this drug isn't even a disease-modifying drug. And so before we sort of assume that it is, I think that we have to design our trials if we are going to do head-on comparison with other, what we believe to be a disease-modifying drug. It may be done in such a way that we prove or disprove that contention. I realize, how hard that would be, but that is possible.

Willkens: Do you think we have a disease-modifying drug?

Furst: I do believe, we have one. I actually think, gold maybe.

Rau: A brief comment to this point: As you heard, 50% of our patients were treated concomitantly with gold and MTX. In these patients our intention was to continue gold as a remittive agent and to give MTX as a very potent anti-inflammatory drug in addition to gold in phases of very high disease activity to reduce the high activity and then to stop MTX again. But in most of these patients gold was stopped and MTX continued, because the patients' impression was that MTX was much more potent than gold.

And now a comment on the patients presented at this symposium: All of them had a long duration of the disease and a far progressed rheumatoid arthritis not only with inflammatory but also with destructive changes of the cartilage and the bone and joint deformities. The normal use of these joints causes pain and enhances the destruction even in the absence of 'rheumatoid' inflammation, that means even if the drug is effective on rheumatoid inflammation. So in cases in which joint destruction is more prominent than joint inflammation it is nearly impossible to test a drug for its disease-modifying properties. For this reason I should propose to take for comparative studies only, or relatively more, patients with less-advanced disease who have more inflammatory rather than destructive joint changes. They would give a much more realistic picture about the antirheumatic properties of the drug.

Now to the mode of application: Dr. *Weinstein,* can you give a short statement on the best way of application: i.v., i.m. or orally as a devided or a single weekly dose.

Weinstein: I don't know that there is a best way. For rheumatoid arthritis, I think, it would be impractical and costly to give it intravenously. Whether or not there is a difference of i.m. versus orally, I think there probably is not a great difference. And I would use it intramuscularly if there are patients who don't respond and you wonder whether or not the bioavailability might be poor in that particular patient. But I don't think there is much to choose otherwise between oral and intramuscular, except for convenience.

Rau: Perhaps Dr. *Michaels* can give a comment on the practicability of intravenous injections.

Michaels: Actually, an intravenous injection is no more difficult than an i.m. injection. In our intravenous study, the patients found it less painful than having their blood drawn.

Rau: One question to Dr. *Furst:* Can you give a short summary of your pharmakokinetic studies and answer the question if in low-dose treatment drug intractions are a problem or not?

Furst: It is my belief based on what Dr. *Jürgens* said, and my review of the literature, that the interaction between nonsteroidals and salicylate and methotrexate is probably very minimal. And even if it existed, I do not believe that we could get around it. Because I do not believe that we can have rheumatoid arthritis patients, who are not on these drugs. So that it may exist, but I don't believe that it's very important. But even if it is, we are stuck with it.

Rau: As far as the contraindications are concerned all of us agree with the rules of the American Academy of Dermatology, published by *Roenigk* and co-workers in 1982.

The same will be true for the laboratory controls during MTX treatment. But one question to Dr. *Willkens:* Do you take chest X-rays of these patients every year?

Willkens: I think that would be advisable.

Williams: I would not take them with the intent of picking up methotrexate-induced pneumonitis, but I will get a chest X-ray on any methotrexate patient who is complaining of shortness of breath. It is so infrequent, and when it occurs, it develops symptoms that I don't think you pick up with a routine chest X-ray, I would not object to routine chest X-rays for other purposes, but I wouldn't do it to pick up methotrexate-induced pneumonitis.

Rau: I would agree that the complication, if it occurs, develops very fast and one cannot expect to catch it when taking routine chest X-rays every year. But these chest X-rays were proposed by the American Academy of Dermatology. Dr. *Weinstein,* would you agree to do without routine chest X-rays?

Weinstein: Yes, I agree. I don't know if you were going to ask this, but I wonder if everyone does a serum-creatinine on the patients before starting them on methotrexate? Do you that Dr. *Rau?* Do you get a serum-creatinine of every single patient that you start on methotrexate in your clinic?

Rau: Yes, serum creatinine is most important and we get it in every single patient that we start on MTX. In cases of elevated serum creatinine the excretion of MTX is delayed, the halflife of the drug is prolonged, the toxicity is increased. If elderly patients get nausea with MTX and for this reason don't drink enough, serum creatinine levels rise, MTX levels rise, and the patient suffers toxicity. So I think it's clear to take renal function tests, liver function tests, the total blood count, the patient shouldn't be pregnant and shouldn't have any infection when starting with MTX. But what do you think about peptic ulcer in the history of the patient? Would be this a contraindication to start MTX or not? In Dr. *Nyfors'* opinion, a history of peptic ulcer is a contraindication of MTX treatment.

Nyfors: If I had a patient, who said he has had gastrointestinal ulcers in the last 10 years, I would X-ray him and see if there was any ulcer before I started, no doubt about that.

Rau: Okay, and the frequency of the controls for instance of the blood count, if the patient is on a low dose for a longer time and he is doing well and had no complication, how often do you think the blood count should be taken, Dr. *Michaels?*

Michaels: What I routinely do, is check the blood count 1 month after they started, as well as the CBC, platelet count and liver function test, and I'll do that for 2 or 3 consecutive months. I have not run into any toxicity earlier, Dr. *Nyfors* suggested that it be checked once a week for the first few weeks. But even with our higher doses we were checking them every week at that point, with a higher dose and ran into no problems. So, I will check it once a month for the first 2 or 3 month, and afterwards I routinely will check the blood work every 2 months. Liver function tests and CBC with platelet count.

Furst: My opinion is a little different. Perhaps based on my personal experiences. But I generally take weekly blood counts for 1 month, every other week for at least 6 months, before I go to 1-month blood counts and liver function tests. My reasoning is that I have had at least 1 patient now, in the last 6 months, whose platelet count dropped within a week from 180,000 or something like that to 90,000. And so I am very careful at the beginning, and then back off very slowly. But I am sure that there is no tried and true method.

Rau: I think the frequency of transaminase elevations depends indeed on the frequency with which you do these investigations. We do it every week, at least in the first 1 or 2 months, and we have elevations in about 30–40% of patients. And if you have elevated SGOT and SGPT, Dr. *Weinstein,* you wouldn't do anything? You wouldn't reduce the dose or stop the treatment, you would continue with the same dose?

Weinstein: Yes, if it was less than three times normal. If it was more than three times normal, I would likely make some minor alterations.

Rau: And Dr. *Williams*?

Williams: One thing that we have not discussed is the time when you have the blood test. There have been some reports to transient elevations immediately after the dose, and I should think that we ought to be sure we're drawing the blood just before their next dose rather than just after the dose. That's not been mentioned and I think that ought to be done first.

Rau: Dr. *Nyfors* told us that investigations should be done 5–7 days after the last dose, that would be before the next dose. That's true for blood count and for liver enzyme values, I suppose. And when would you see the indication for a liver biopsy. Would you follow the recommendations of the American Academy of Dermatology or not, that means, would you recommend a biopsy before starting a treatment? Dr. *Nyfors* would recommend, Dr. *Willkens,* what's your opinion?

Willkens: I am not prepared to answer to that yet. I'm doing a prospective study and will be able to answer that when I have my data in hand. It's not been my practice to do a pretreatment biopsy, and I suspect that we are going to come up with some decision with respect to the duration of time the patient has been on the drug, whether or not he has significantly elevated liver enzyme abnormalities, and/or has taken more than a certain amount of drug. It may be 1.5 g, it may be more than that. But I think that we have to be prepared for some time to do some sort of hepatocellular evaluation independent of liver enzyme changes.

Rau: So, you would say, now you don't do pretreatment biopsies, but you will be able to answer the question after what cumulative dose or under what conditions a biopsy should be done?

Willkens: That's right. I will be, but (I am) not now.

Rau: And I should ask Dr. *Williams* to sum up briefly the question of lung complications.

Williams: There appears to be methotrexate-induced pneumonitis that occurs infrequently and resolves relatively spontaneously. The etiology and pathogenesis is unknown.

Rau: Okay, thank you.

Concluding Remarks

In an attempt to sum up the symposium the following conclusions can be drawn.

(1) MTX given in a single or divided weekly low dose is a very *potent and effective drug* in the treatment of psoriasis, psoriatic arthritis and rheumatoid arthritis. The therapeutic effect on clinical and laboratory data (for instance morning stiffness, joint count, grip strength, daily needed steroid dose, ESR, acute-phase reactants) can be observed already after 4–6 weeks of treatment, that is earlier than with any other cytostatic compound or slow-acting antirheumatic drug. As far as the extra-articular manifestations of rheumatoid arthritis are concerned, there seems to be no effect on the development of rheumatoid nodules; on the other hand, a steroid-sparing effect in the treatment of vasculitis and pericardial effusion can be observed.

(2) The *mechanism of action* of MTX in rheumatoid arthritis is not fully understood: Experimental data and the rapid clinical efficacy as well as the rapid flare up after discontinuation of treatment support the thesis of a direct anti-inflammatory effect. But according to experimental data and to the observation of herpes zoster virus infection in patients treated with low dose MTX, immunosuppressive mechanisms seem to be involved, although clinical trials could not demonstrate any immunological changes in patients successfully treated with MTX. So we cannot decide whether MTX should be understood as a merely symptomatic agent like steroids or as an agent that perhaps influences the fundamental underlying mechanisms of rheumatoid arthritis as for instance gold salts are supposed to do. The term 'remission-inducing' or 'disease-modifying' drug has been questioned, because one has to continue the treatment to sustain the effect. Even after discontinuation of gold treatment a relapse can occur. Up to date nothing is known about the radiological progression, that means the development of erosions and joint destructions, in patients on long-term treatment

with MTX. Studies in this direction should be undertaken and they are underway.

(3) Low dose MTX is *indicated* in all very active RA cases not responding to conventional treatment with antimalarials, gold or *D*-penicillamine. On the other hand, in severe rapidly progressing so-called malignant diseases we should not wait until a full course of gold and after this *D*-penicillamine has been given. MTX is a true alternative to gold or *D*-penicillamine. It is an open question whether MTX can be combined with other slow-acting antirheumatic drugs. In about 50% of cases we prescribed MTX in addition to gold salts in very active cases to get a rapid effect; it was our intention to stop MTX again when gold salts were working. We do not have the impression that this combination is laden with more side effects than MTX or gold salts alone, but we do not yet know if both drugs have an additional or potentiating ameliorating effect on rheumatoid arthritis. To answer this question further investigations should be done. At the Cleveland Clinic the patients are regularly treated with a combination of MTX and antimalarials.

(4) As far as *contraindications* are concerned, most of the authors seem to follow the recommendations of the American Academy of Dermatology: The prescription of MTX is contraindicated in cases of renal function disturbances, significant liver function abnormalities, pregnancy, severe bone marrow disturbances, active peptic ulcer, excessive alcohol consumption, active infectious diseases, adipositas and in unreliable patients.

(5) Before starting the treatment, the following *laboratory studies* should be done: total blood count with differential count and platelet count; urinalysis and serum creatinine or creatinine clearance; routine liver function tests, especially transaminases; liver biopsy (?); X-ray film of the chest (?). During treatment, blood count should be studied in 1–4 weekly intervals, urinalysis, serum creatinine and liver enzymes in intervals of 2–3 months. Corresponding with *Furst* we prefer to take laboratory data including total blood count, serum creatinine and liver function tests every week during the first 2 months of treatment and every 1–2 months thereafter.

(6) When these precautions are followed and regular controls are made, MTX is a relatively save agent comparing favorably with other cytostatic drugs or even with gold salts, the rate of withdrawals because of side effects being very small. The most common *side effects*

are nausea and vomitus in up to 50% of patients, oral discomfort in 20% and stomatitis in 6% of patients; loss of hair, diarrhea and fatigue are less frequently observed. In most cases these conditions disappear after dose reduction. Bone marrow or pulmonary complications are very uncommon and, in contrast to other cytotoxic agents, a carcinogenetic potential can be neglected.

Hepatotoxicity is the most difficult long-term problem in patients treated with MTX. The discussion about this side effect remains controversial. In patients with psoriasis, cases of fibrosis and cirrhosis have been reported. *Nyfors* described piecemeal necrosis which progressed to cirrhosis. Other investigators were unable to find statistical differences in respect to liver changes between treated and untreated patients with psoriasis. In rheumatoid arthritis, *Mackenzie* and *Frenzel*, respectively, found the same rate of inflammatory infiltrates in the portal fields and the same alterations of liver cells in patients treated with MTX as in untreated RA patients. Moreover, in *Mackenzie's* series there were no differences between groups with different cumulative doses of MTX. Alcohol consumption, obesity, diabetes, old age seem to be additional risk factors. Transient elevations of liver enzymes can be observed in up to 40% of patients, but as they are relatively useless in predicting the development of cirrhosis or fibrosis, elevations up to 3 times normal can be disregarded, above this level an adjustment in MTX dose is advisable. The most reliable method to detect structural changes of the liver remains liver biopsy. Indications for a biopsy in MTX-treated psoriasis patients are given by *Roenigk* et al. [1972, cf. p. 216]. For rheumatoid patients, recommendations can be given only after the results of prospective studies are available.

(7) *Drug interactions* with salicylates or other nonsteroidals do not play a major role in low dose MTX treatment in case of a normal renal function, for according to *Jürgens* at low doses of MTX tubular secretion of MTX does not play a major role. So one could not expect a high degree of drug interaction between MTX and weak acids, competing with MTX for the active membrane carrier of tubular secretion. On the other hand, there are reports showing drug interactions between low dose MTX and nonsteroidal anti-inflammatory drugs as well as steroids. Therefore, according to *Furst*, we have to consider that the pharmacokinetics of low dose MTX have to be further evaluated.

(8) Between the *oral, intramuscular* or *intravenous* mode of application there seem to be no differences in respect of bioavailability,

efficacy and toxicity, as oral doses up to 30 mg are completely absorbed. But according to *Furst* there are differences in absorption between individual patients even with low doses of MTX. It has to be further evaluated if there are differences between the single weekly and the three divided weekly oral dose.

R. Rau

Subject Index